ENGINEERING
AS A
CAREER

ENGINEERING AS A CAREER

FOURTH EDITION

Ralph J. Smith
Stanford University

Blaine R. Butler
Purdue University

William K. LeBold
Purdue University

McGRAW-HILL BOOK COMPANY

New York St. Louis San Francisco Auckland Bogotá
Hamburg Johannesburg London Madrid Mexico Montreal New Delhi
Panama Paris São Paulo Singapore Sydney Tokyo Toronto

TO ANN, DONNA, LOUISE, AND CINDY

In appreciation of their encouragement,
suggestions, and assistance over the
past two years.

This book was set in Optima by A Graphic Method Inc.
The editors were Julienne V. Brown, Kiran Verma, and J. W. Maisel;
the cover was designed by Irving Freeman;
the production supervisor was Leroy A. Young.
New drawings were done by Danmark & Michaels, Inc.
Kingsport Press, Inc., was printer and binder.

ENGINEERING AS A CAREER

3 4 5 6 7 8 9 0   KGPKGP   8 9 8 7 6 5

ISBN 0-07-058788-4

Library of Congress Cataloging in Publication Data

Smith, Ralph Judson.
    Engineering as a career.

    Includes bibliographies and index.
    1. Engineering—Vocational guidance. I. Butler,
Blaine R.   II. LeBold, William Kerns, date
III. Title.
TA157.S6 1983   620'.0023   82-16193
ISBN 0-07-058788-4

CONTENTS

PREFACE

The role of the engineer in our society has expanded dramatically in the last decade. We face major problems in supplying energy, protecting the environment, providing safer and more efficient transportation, and increasing productivity. We must improve communication, speed processing of information, and facilitate solving problems of increasing complexity. We need to apply technological advances in improving the quality and quantity of food and water and the effectiveness of medical diagnosis and health care. We have to solve the problems of inner cities while we learn more about outer space.

For the solution of these and other pressing problems, society will look to the engineering profession. We must attract to this challenging profession gifted young men and women with the intellectual abilities and personal qualifications to contribute to the improvement of the quality of life in a rapidly changing world.

However, our ability to provide professional education of high quality is limited by engineering faculty availability, laboratory capacity, and data-processing facilities. For the next decade, the number of freshmen enrolling in engineering will far exceed the number of spaces available for seniors. Vocational orientation and career counseling are especially important now.

In this fourth edition, the basic purpose of the book remains unchanged: to acquaint beginning students with engineering as a career. The revision provides an opportunity to catch up with the significant advances of the past ten years and to incorporate the many helpful suggestions of those who have used the textbook in freshman orientation courses.

Objectives: It is hoped that these changes will result in an improved book to accomplish the following four major objectives:

To answer the frequently asked questions: What do engineers do? What qualifications do they need. Ideally, students considering engineering should compare their abilities and interests with the qualifications needed for success in this exacting profession. Many students actually begin engineering study without adequate information only to discover several years later that engineering is not what they imagined. One aim of this book is to present a factual picture of engineering activities and responsibilities and the corresponding aptitudes and training. An accurate and up-to-date description should aid in vocational counseling and should result in an improved engineering profession.

To characterize the engineering profession in terms of functions as well as branches. Engineering can be classified by *branch* (civil, electrical, mechanical, chemical, aeronautical, industrial) or by *function* (research, development, design, construction, production, operation, application, management). A student's *interest* may lie in a certain branch but his or her *aptitudes* are most likely to be related to a certain function. The functional classification is therefore much more meaningful in career planning. This emphasis on the functional classification has proved highly valuable to previous users of the book, so it has been retained, with many new examples.

To portray the larger role of the engineer with emphasis on creativity and decision making as well as analysis and design. Creativity, a key attribute of successful designers, is illustrated by examples drawn from various fields. The environment required for creativity is discussed, and the creative process is described in detail. How this relates to an individual and how an engineering student can take advantage of what we know about creativity are then discussed. Decision making, an essential step in the design process, is discussed in terms of the place of the engineer in society and the effect of technical decisions on human welfare. A new chapter is devoted to the challenges facing engineers in the areas of energy, pollution, space travel, and productivity. These illustrations provide a basis for discussion of the future role of the engineer in a technological society.

To provide assistance in learning how to study, how to write reports, and how to solve engineering problems. Successful adjustment to academic life is a prerequisite to academic achievement. Practical help is provided in the form of general principles of effective learning and specific suggestions that have proved valuable in improving study habits, increasing reading speed, and improving performance in examinations. In response to requests, a chapter on writing and presenting reports has been included. The importance of skill in verbal communication is emphasized and an efficient procedure for writing effective reports is outlined. The activity characteristic of all engineering is problem solving, but the solution of complete problems is usually deferred until late in the college curriculum. To provide motivation for study of the engineering sciences and to give students a feel for the relation between knowledge and judgement in engineering practice, some experience in problem solving is included here.

Contents. The contents are organized into four parts. The first four chapters provide a *description of the engineering profession.* After a motivating introduction to career selection, students learn six key characteristics defining engineering, gain perspective from a brief history of technological achievements, and are introduced to technical societies and professional registration.

The next five chapters describe in detail *the nature of engineering.* Students see the classification of engineering into major and minor branches and the activities associated with each. Then they learn the functional classification and how aptitudes and training are related to the function performed is considered. Through interesting examples, they come to understand the importance of creativity and decision making in modern engineering design. Then they study the relations between engineers, scientists, and technicians and the ways that members of a technical team differ in their preparation and their responsibilities. Finally, they obtain an overall picture of the engineering profession through statistics on past growth, current distribution by branch and function, factors affecting earnings, and future employment outlook.

The third part is aimed at *college preparation for a career,* with emphasis on material that freshmen need but do not ordinarily get in engineering courses. After being introduced to the principles of learning, students learn how to study, how to benefit from lectures, and how to improve their reading ability. Then they see why report writing is important and they learn to "design" and write a report. Finally, they gain insight into the philosophy and technique of problem solving and learn a general method that can be applied to problems in many fields.

The last part provides students with a look at their *future in engineering,* starting with a stimulating discussion of challenging problems that remain to be solved. In the final chapter they draw conclusions about engineering as a possible career and face the questions that must be answered affirmatively by students truly committed to this exciting, but demanding, profession.

This book is designed to provide the orientation portion of an introductory engineering course. The text material can be covered in about 15 or 20 lectures. Most frequently it is used in conjunction with other activities, such as a series of lectures by practicing engineers, a program of field trips, or a course in computation. By providing a complete and consistent treatment of branches and functions, the textbook allows specialists to concentrate their time and effort on specific topics without the need to include full detail or total balance.

An Instructor's Manual is available upon request to teachers using this text. Planned as an aid in making each hour of the instructor's time as productive as possible, the manual includes suggestions for course and class organization, supplementary information on topics of current and specialized interest, illustrative anecdotes and quotations, and solutions to text problems.

This edition is primarily the work of coauthors Blaine Butler and William LeBold. Taking advantage of their experience with the highly regarded freshman engineering program at Purdue University, they have brought a new perspective to this work. They have revised the approach, up-dated the statistics and examples, and modernized the content. I am pleased to acknowledge their contributions.

Ralph J. Smith

ENGINEERING
AS A
CAREER

Kennedy Space Center, Florida. The space shuttle Columbia rises off pad 39A a few seconds past 7 A.M. on April 12, 1981—the dawn of a new era of engineering challenge and achievement. (*NASA*)

The nature of engineering
  Opportunities in engineering
 Qualifications for success
A look to the future

1

INTRODUCTION: ENGINEERING AS A CAREER

It is dawn, and silhouetted against the lightening sky stands one of the most sophisticated creations of the engineer's art. The countdown has begun, and a hundred technicians are monitoring the performance of the multitude of components that make up the space shuttle. With a flash and a roar the three main engines and the two parallel-burning solid-rocket booster engines lift the huge vehicle slowly from the pad. Rising ever more swiftly, it climbs into the sky. Two minutes after lift-off, twenty-five miles down-range at an altitude of 140,000 feet, the solid rocket boosters separate and fall into the sea. The space shuttle continues to ascend. Eight minutes after lift-off, the main engine cut-off occurs at an altitude of 380,000 feet at a velocity of 25,574 feet per second. The external tank separates, the orbital maneuvering engines start, and the space shuttle is eased into its circular parking orbit 150 nautical miles above the earth. Twelve hours later, after all systems have been checked, the space tug and its attached communication satellite are lifted from the cargo bay. The space tug engines ignite. Then, the space tug ascends to its programmed altitude and velocity, and releases the communication satellite into its planned orbit. Solar panels are swung out to

1

capture energy from the sun, antennas are extended, and transmitters and receivers are turned on.

As it silently orbits around the earth, the satellite passes over vast agricultural areas that once were dry deserts and over broad rivers that have been converted from destructive torrents to provide water, power, transportation, and recreation. It looks down on great cities by the water's edge; within the cities, commerce flourishes, and on the outskirts, factories pour out a stream of products. Millions of healthy, well-fed, educated citizens live, work, and play, enjoying a standard of living unmatched in any former era.

A closer look reveals some striking contrasts. Over some of those great cities hang the brown clouds of pollution, and in the crowded inner cities none of the human benefits of technological progress is evident. The contrails of high-flying jets mark the international air lanes, but below, a peasant dispiritedly follows his weary bullock as it drags a wooden plow through depleted soil. While the affluent nations are surrounded by plenty, millions of less fortunate people, unable to cope with nature, live under the threat of famine, drought, and pestilence. While citizens of the advanced nations are in constant communication with the news centers of the world, the people of backward areas live in ignorance and superstition, unaware of what is happening in the next village. While the face of the earth is being transformed by engineering, much more remains to be done.

- What is engineering?
- What do engineers do?
- What types of engineering careers are available?
- What does engineering have to offer as a lifetime vocation and professional career?
- What qualifications and interests are required for success and satisfaction in engineering?

This book has been written to provide answers to these and other questions in the minds of persons considering a career in professional engineering.

The selection of a career is one of the most important decisions you will ever make, and it is a decision that you must make yourself. You can follow a trial-and-error process, trying one type of work after another until you are satisfied, or you can proceed systematically, attempting to match your interests and aptitudes with the opportunities and requirements of available occupations and careers.

Career selection can be broken down into three steps. First, you could decide what you want out of life—material success, opportunity for service, security, freedom of action, opportunity for leadership, challenge, social status, prestige, personal privacy, creative satisfaction, or freedom from pressure. On these and other points you could decide what will contribute most to the life you wish to lead. Second, you could decide what aptitudes and other qualifications you possess. Are you skillful in working with ideas, with things, with mathematics, with equipment, with people, with money? Are you artistic, scientific, industrious, practical, creative, socially conscious, enterprising? What are your weaknesses? Do you have limitations in your problem solving ability, mechanical ability, oral and written communications skills, spatial visualization, and human relations skills?

Having determined what you are after and what qualifications you have, you could then consider a variety of careers and see which one comes closest to offering what you want in return for work for which you are qualified. If this book helps you to understand the nature of engineering, the opportunities for interesting work and satisfying achievement, and the mental and personal qualifications required of engineers, then its purpose will be accomplished.

THE NATURE OF ENGINEERING

In many respects the story of engineering is the story of civilization itself.[1] Civilization has been described as the process whereby human beings strive to overcome obstacles and become master of the environment. Engineers are constantly at work seeking to understand the laws of the physical universe, to take advantage of the beneficial aspects of nature, and to convert or modify the adverse factors. While overcoming the material obstacles faced by men and women, the engineer also contributes to the moral and spiritual welfare of people by removing some of the causes of want and fear. Sometimes, engineers are viewed as antagonists of existentialists, but the authors believe that creative productive engineers are in many ways existentialists. They reject dogmas that say scientific-technological solutions are not desirable or possible and rely on the impulses and intentions that are the grounds for existence in an engineering sense.[2]

Perhaps you were attracted to the study of engineering without knowing just what it is. In describing the engineering profession, we shall first set up working definitions of engineering and engineer and later on distinguish between engineers and their coworkers, the scientist and technologist. To provide the proper perspective for studying engineering as it is practiced today, a brief history of engineering from ancient to modern times is included. It will be seen that engineering, like civilization, is a changing thing, and the rate of change is ever-increasing.

The primary emphasis in the first half of the book, however, is on what engineers do and how they do it. The broad field will be discussed in terms

[1] James K. Finch, *Engineering and Western Civilization*, McGraw-Hill, New York, 1951.
[2] Samuel E. Florman, *The Existential Pleasures of Engineering*, St. Martins, New York, 1976.

of the branches of civil, mechanical, electrical, materials, aerospace, industrial, and chemical engineering and the more specialized subbranches. Of particular interest to you as a beginning student are the functions of engineering—research, development, design, construction, production, operation, sales, and management. You will be able to compare your aptitudes with those required for success in each of the various functions.

OPPORTUNITIES IN ENGINEERING

Engineering is one of the largest of all professions, with nearly 1.5 million members in the United States. Although it is the youngest of the organized professions, its record of achievement is already impressive. Public recognition of the contributions made by engineers has brought new opportunities for service and, at the same time, an even greater need for high standards of personal and professional conduct (see Chapter 4).

Is engineering a rewarding profession? The opportunity for material rewards for engineering work appears to be quite good. Of greater importance, however, is the sense of accomplishment that comes with creative and constructive work. Engineers are inventors, creators, designers, builders, operators, and managers. As an engineer, you will play an important role in the development of our technological society. To make the optimal contribution, you will need to understand how our society operates and how individuals behave; in this connection the necessity for courses in communications, social sciences, and humanities will become clear.

QUALIFICATIONS FOR SUCCESS IN ENGINEERING

Are you suited for work in engineering? Engineering is not a single activity; rather, it is a broad spectrum of activities or functions. While all functions of engineering have some common characteristics, there is a wide variation in emphasis, and certain abilities are of much greater importance in some engineering activities than in others. A major section of this book is devoted to a detailed discussion of desirable mental and personal qualifications to help you decide whether engineering is the career for you and where in engineering you would fit best.

Much of the engineer's work is concerned with the design of structures, devices, processes, and systems to serve human needs. An important factor in this ability to satisfy needs is a person's creativity. In Chapter 7, some examples of creativity are presented and the creative process is described to give you some ideas about developing your own creativity. Another important attribute for the engineer is his or her ability to make decisions. In addition, engineering is related to some important problems involving social factors so that you can see how technical decisions influence the human environment and whether this responsibility is appealing to you.

But having the technical ability to solve society's problems is not enough. The engineer can build nuclear power plants, but is it politically and economically feasible to do so? Much needed oil can be extracted from oil shale, but is it environmentally acceptable to do so? One-half of the world's known coal reserves are located in the United States, but are the costs of pollution and acid rain too high to justify expanding our use of coal?

Many of the technical problems that face our country today require political and social consensus before action will be undertaken. Engineers must participate in these decisions and must be aware of social and political

A robotic arm, developed by mechanical engineers and rehabilitation experts, offers a piece of chocolate to a patient. A microprocessor in the robotic arm is programmed to perform a variety of manual tasks for disabled patients. To satisfy human needs, engineers frequently work with experts in other fields, such as medical doctors, to apply science to solving real problems.

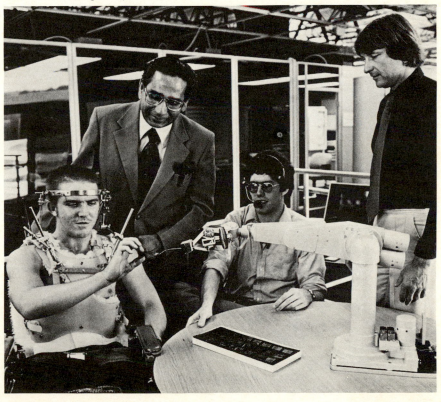

implications. They must help to insure that the public understands the trade-offs in various solutions so a consensus can be reached and a publicly acceptable technical solution can be completed. Chapter 13 outlines some of the major problems that face our country today.

A LOOK TO THE FUTURE

Do you want to change the world? In our technological society the engineer is one of the most powerful agents for change. Engineers are finding new ways to control our environment, to utilize the forces of nature, to increase our productivity, and to improve the quality of life for all. The engineers who will, in large part, be responsible for planning the world of the year 2000 are just starting their professional training. The power they will hold makes engineering an exciting prospect, as well as a sobering responsibility. The easy problems are always solved first; those remaining are bound to be difficult. The need for well qualified persons is obvious; you are embarking on a lifetime of challenge if you decide to choose engineering as a career.

Some proposed definitions
 Essential characteristics of engineering
Characteristics of the engineer

2

DEFINITION OF
THE ENGINEERING
PROFESSION

There is a great deal of confusion in the minds of the general public as to just what engineering is and what men and women with engineering educations do. To remove some of that confusion, let us take time now to set up some working definitions to be used throughout the book. We shall first examine definitions that have been proposed from time to time and pick out what appear to be the fundamental elements of a satisfactory definition of engineering and engineer. Subsequent chapters will provide illustrations of engineering projects and detailed discussions of the work of engineers in the various functions they perform to make clear the nature of engineering.

SOME PROPOSED DEFINITIONS

As a profession, engineering is relatively new, and engineering activities are continually changing in nature and scope. Among the many attempts that have been made to define engineering, the following, listed in chronological order, are of interest. As you read them, note the varying emphasis in the statements.

A. M. Wellington (1887): It would be well if engineering were less generally thought of . . . as the art of constructing. In a certain important sense it is rather the art of not constructing . . . of doing that well with one dollar which any bungler can do with two after a fashion.

S. E. Lindsay (1920): Engineering is the practice of safe and economic application of the scientific laws governing the forces and materials of nature by means of organization, design, and construction, for the general benefit of mankind.

J. A. L. Waddell, Frank W. Skinner, and H. E. Wessman (1933): Engineering is the science and art of efficient dealing with materials and forces . . . it involves the most economic design and execution . . . assuring, when properly performed, the most advantageous combination of accuracy, safety, durability, speed, simplicity, efficiency, and economy possible for the conditions of design and service.

Vannevar Bush (1939): Engineering . . . in a broad sense . . . is applying science in an economic manner to the needs of mankind.

T.J. Hoover and J. C. L. Fish (1941): Engineering is the professional and systematic application of science to the efficient utilization of natural resources to produce wealth.

L. M. K. Boelter (1957): Engineers participate in the activities which make the resources of nature available in a form beneficial to man and provide systems which will perform optimally and economically.

John C. Calhoun, Jr. (1963): It is the engineer's responsibility to be aware of social needs and to decide how the laws of science can be best adapted through engineering works to fulfill those needs.

The Accreditation Board for Engineering and Technology (1982): Engineering is the profession in which a knowledge of the mathematical and natural sciences, gained by study, experience, and practice, is applied with judgment to develop ways to utilize, economically, the materials and forces of nature for the benefit of mankind.

There also are two significant definitions of engineer. The first is important because of its recognition as law. As a basis for professional registration, each state must provide a legal definition of engineer and the practice of engineering. A typical provision[1] is as follows:

"Professional engineer," within the meaning and intent of this act, refers to a person engaged in professional practice of rendering service or creative work requiring education, training and experience in engineering sciences and the application of special knowledge of the mathematical, physical, and engineering sciences in such professional or creative work as consultation, investigation,

[1]The Civil and Professional Engineers Act of the State of California, Art. 1, chap. 7, div. 3, Business and Professions Code.

evaluation, planning or design of public or private utilities, structures, machines, processes, circuits, buildings, equipment or projects, and supervision of construction for the purpose of securing compliance with specifications and design for any such work.

The second definition is significant because it has been approved by representatives of a large segment of the engineering profession. Accepted by delegates to a conference of engineering societies of the United States and Western Europe is the following:

A professional engineer is competent by virtue of his fundamental education and training to apply the scientific method and outlook to the solution of problems and to assume personal responsibility for the development and application of engineering science and techniques especially in research, designing, manufacturing, superintending, and managing.

ESSENTIAL CHARACTERISTICS OF ENGINEERING

As we study the foregoing definitions, certain key words and phrases appear over and over again—art, science, application of science, organizing and directing, sources of power, natural resources, safety, economy, efficiency, professional, scientific method, benefit of the human race, good of humanity. A careful consideration of these definitions and engineering as it is practiced today leads to the conclusion that there are six essential elements in defining engineering:

Engineering is a profession.
Engineering is an art rather than a science.
Engineering is based on the application of science.
Engineering is concerned with optimum efficiency, economy, and safety.
Engineering involves the utilization of natural resources.
Engineering has as its ultimate purpose the benefit of mankind.

Incorporating these six elements into a single statement results in the following:

Engineering is the professional art of applying science to the optimum conversion of natural resources for the benefit of mankind.

This definition has the virtue of brevity, but its use requires a thorough understanding of the key phrases. Let us consider the six elements, one at a time.

Professional

What is a profession? The vocations commonly referred to as professions include law, medicine, ministry, teaching, architecture, and engineering.

The vocations have four common characteristics:

Associated with a profession is a great body of special knowledge.

Preparation for a profession includes training in applying that knowledge.

The standards of a profession are maintained at a high level through the force of organization or concerted opinion.

Each member of a profession recognizes his or her responsibilities to the public over and above responsibilities to clients or to other members of the profession.

Certainly engineering qualifies as a profession on all four counts.

Art

An art calls for the systematic application of knowledge and skill according to a set of rules. Engineering requires ingenuity, craftsmanship, and judgment in adapting knowledge to practical purposes. A major engineering activity is problem solving, and for some engineers this is an art. The engineering method of problem solution includes a clear statement of the problem and necessary assumptions, the mental creation of a concept or device or system that appears to meet the need, a logical analysis of the situation based on established principles, a careful checking of the results, and a set of conclusions or recommendations based on the facts. On the basis of inadequate information, the engineer must design for the unknown future. The ability to conceive an original solution and predict its performance and cost is a distinguishing attribute of the professional engineer.

Applying Science

Science is accumulated, established, and systematized knowledge. Engineering is based on the fundamental sciences of physics, chemistry, and mathematics and their extension into materials science, mechanics, thermodynamics, electrodynamics, and transfer processes. The word "science" is derived from the Latin word *scire,* which means "to know." In contrast, the function of the engineer is "to do." The scientist is concerned with adding to our store of knowledge. The engineer brings science to bear on practical problems; the engineer is a person of action.

The engineer uses science, but is not limited to scientific knowledge. No one knows exactly how and why concrete and steel behave as they do, but, using empirical data, the engineer is able to design efficient and safe structures. It is important to note that the conception and design of a structure, device, or system to meet specified conditions in an optimum manner is engineering even when it is done by a person whose formal training is in science, or when it is for the purpose of gaining new knowledge. Frequently

Soaring 630 feet into space, the Gateway Arch on the Mississippi River at St. Louis is an artistic and technological wonder. Designed by Eero Saarinen, this lean steel structure memorializes the spirit of the pioneers who settled the West as it follows the curve of an inverted catenary. (A uniform cable supported at the ends hangs in the catenary $y = k \cosh x/k - k$ where the constant k depends on the cable weight per foot and the tension.)

Great structural strength is obtained by the stressed-skin design similar to that used in aircraft; the observation deck at the apex would sway less than 18 inches in a 150-mile per hour wind. How would you build such a structure? The engineer solved this problem by using creeper cranes mounted on tracks fixed to each leg of the arch. After the final section was placed, the cranes crept back down dismantling their tracks as they came.

For those who wish to emulate the hardy pioneers whose spirit the arch commemorates, there is a climb of 1076 steps from the base to the observation deck. Less ambitious visitors are carried to the top in trains of five-passenger capsules. To provide a level ride inside the curved arch, the engineers mounted the capsules like the baskets of a Ferris wheel. (*The Chamber of Commerce of Metropolitan St. Louis*)

the popular press adds to the confusion by hailing an orbiting satellite as a "scientific achievement" and describing an unsuccessful launching as an "engineering failure."

Optimum Conversion

Unlike scientists, engineers are not free to select the problems that interest them; they must solve problems as they arise, and their solutions must satisfy conflicting requirements. Efficiency costs money, safety adds complexity, performance increases weight. The engineering solution is the optimum solution, the most desirable end result taking into account many factors. It may be the cheapest for a given performance, the most reliable for a given weight, the simplest for a given safety, or the most efficient for a given cost.

To the engineer, efficiency means output divided by input. The job is to secure a maximum output for a given input or to secure a given output with a minimum input. The ratio may be expressed in terms of energy, materials, money, time, or human power. Most commonly the denominator is money; in fact, most engineering problems are answered ultimately in dollars and cents. Efficient conversion is accomplished by using efficient methods, devices, and organizations.

The emphasis on efficiency leads to the large, complex operations characteristic of engineering. The processing of the new antibiotics and vaccines in the test-tube stage belongs in the field of biochemistry, but when great quantities must be produced at low cost, it becomes an engineering problem. It is the desire for efficiency and economy that differentiates ceramic engineering from the work of the potter, textile engineering from weaving, and agricultural engineering from farming.

With many large engineering problems the social costs are an important consideration. Will a new freeway require the removal of homes and the relocation of people in unfamiliar surroundings? Or, will the failure to build a freeway result in lack of access to an area, with resulting industrial stagnation and loss of jobs? Will a computerized production line utilizing robots throw humans out of work? Or, instead, will this change increase the productivity of the workers' industry, keep the particular product competitive in the world market, and increase the workers' standard of living?

Natural Resources

In general there are two types of natural resources—materials and energies. These are the elements from which engineers have fashioned their works.

Engineering materials include animal, vegetable, and mineral items—some are natural but most are manufactured. These materials are useful because of their properties: their strength, ease of fabrication, lightness, or

An offshore mobile oil rig in the North Sea extends the search for natural resources to the continental shelf. Planning multimillion dollar projects of this nature requires teams of engineers and scientists. Mechanical and civil engineers design these unique floating platforms, aerodynamicists estimate worst-case wind loads on the superstructure, and control engineers plan systems to stabilize them in hurricane weather. Petroleum engineers and geologists contribute their special knowledge of drilling technology as key members of the design team. *(Exxon)*

durability, their ability to insulate or conduct, and their thermal, chemical, electrical, or acoustical characteristics. A list of useful materials would be almost limitless; for example, there are 45 metallic elements and approximately 10,000 metallic alloys in commercial use today. By varying the composition of an alloy, the engineer can improve its strength, machinability, conductivity, corrosion resistance, magnetic properties, or bearing characteristics.

The number of important sources of energy is much smaller—coal, petroleum, gas, wind, sunlight, falling water, nuclear fission. Each energy form has advantages and disadvantages. Coal is cheap, but mining it is dangerous and its sulfur content is expensive to remove. Petroleum products can be stored and converted into heat under carefully controlled conditions. The world supply of petroleum is rapidly being depleted and its availability is subject to the whims of foreign powers. Wind power is cheap but undependable. Development of waterpower is feasible only in certain, usually remote, areas. Nuclear fuel is cheap but the conversion equipment is expensive and society is concerned about its safety. Each day, there comes to the earth as sunshine more than 10,000 times as much energy as humankind uses for all purposes. However, we have not yet found ways to effectively convert this solar energy in a way that is competitive with other forms of energy.

Some natural resources are renewable; others may soon be exhausted. The engineer must be concerned with conservation, which means more than just "not using." True conservation requires continual development of new resources as well as efficient utilization of present ones. We are all aware of rapid depletion of such resources as petroleum and high-grade iron ore. In view of the rapid population growth, the desire for a rising standard of living, and the increasing per capita use of material and energy, the conservation of natural resources has become a primary activity of the engineer. The history of civilization seems to indicate, however, that long before our present resources are exhausted, engineers and scientists will have uncovered new sources, developed improved methods of processing, and discovered better replacements.

Benefit of Mankind

In the past we assumed that a product that satisfied some material want was a benefit. An engineer with this attitude might receive complete satisfaction from the skillful solution of a technical problem. Now we are becoming increasingly aware that the contributions of the engineer have political, social, and aesthetic implications far beyond the immediate technical results. An intercontinental missile system may bring security to one segment of the human society while threatening another. A dam to harness the energy of a river in a remote canyon may flood a wilderness area that plays an important role in people's total existence. An efficient and cheap automobile engine that reduces the energy problem may spoil the environment by polluting the air.

The various registration acts justify the legal recognition of the engineering profession: "in order to safeguard life, health, property, and public welfare." Certainly all engineering work must meet the test: Will it be of benefit to human beings?

CHARACTERISTICS OF THE ENGINEER

Having developed a working definition of engineering, we have a basis for prescribing what constitutes an engineer.

> An engineer is a person qualified by aptitude, education, and experience to perform engineering functions.

> Let us analyze the key phrases in this definition.

Aptitude

Aptitude implies a disposition toward or a capacity for work in a certain field. One characteristic of engineering study is that a certain type of mind is desirable. The purpose of engineering education is to draw out, develop, and train inherent abilities or aptitudes.

The contribution of the engineer to the quality of life is most striking in countries that have not participated in the benefits of advanced technology. It has been estimated that one-half of the people in India go to bed hungry each night; a major cause of this is poor soil depleted by centuries of farming with no replenishment of nutrients. This first truckload of fertilizer from a new plant owned and operated by Indians signals the beginning of a new era for this region. With an improvement in agricultural productivity will come an increase in food supply, a reduction in disease, an improvement in human productivity, a reduction in unemployment, and an uplift of hope.

It is, of course, not possible to say with certainty whether or not a particular man or woman should enter the engineering profession; however, certain traits are closely correlated with success in engineering. These include interest in and aptitude for the basic sciences of mathematics, physics, and chemistry; skill in applying science to practical problems; ability to visualize physical relations described in words; facility in translating verbal statements into mathematical terms and interpreting mathematical results in terms of practical objectives.

Engineering is essentially a mental activity. While the engineer is concerned with machines, structures, and other concrete objects, his or her particular responsibility is the solution of abstract problems involving design, performance, efficiency, and cost. Sometimes a liking for taking clocks apart or building model airplanes is incorrectly presumed to be positive evidence of aptitude for engineering. Skill in manipulating tools and machines is more important to the technologist than to the engineer, although practical knowledge of the physical characteristics of technical devices is valuable as background for engineering study.

Education

The development of engineering education has kept pace with the progress of engineering knowledge. A good engineering school provides an ideal environment in which young engineers can, in a relatively short time, prepare themselves for a great variety of starting positions.

Early engineers had little or no formal training. The body of recorded technical knowledge was limited, and it was common practice to serve an apprenticeship under an engineer of some reputation. Information picked up on the job was supplemented by self-education and occasional opportunities for formal schooling.

As the results of experience and research replaced trial-and-error methods, the established engineering principles became so numerous that no one individual could provide the necessary training. A program of systematic instruction in an organized engineering school became the accepted road to professional practice. Recent surveys of engineers indicate that over 95% of the engineers now practicing have completed a college program; the figures range from over 96% of those under 25 years old to 83% of those over 55 years old. In 1930 the overall figure was only 73%.

The engineering faculty is a selected corps of specialists in their various fields. Along with the laboratories and libraries, the faculty offers stimulating opportunities for learning that are not to be found in any other situation. The typical engineering curriculum is a carefully designed and integrated program providing an intensive, efficient, challenging, and even pleasant period of education.

Experience

Engineers are not completely trained in college; graduation is not the end of training. For efficiency the college program must be quite general, emphasizing fundamental principles and omitting the specialized knowledge needed on a given job. In most states the college graduate is permitted to take, as a first step toward professional registration, an examination covering engineering principles and application techniques. Those successful in the examination are certified as Engineers-in-Training (E.I.T.).

Professional engineering, however, is an art requiring practice in its development. Experience is needed to develop judgment as a basis for decisions. The important engineering problems cannot be reduced to simple formulas, nor can answers be found in charts and tables. Engineers are valuable when they are able to make sound decisions based on scientific principles in combination with practical experience. That ability is tested by a second examination leading to registration as a Professional Engineer (P.E.).

Engineering Functions   Although engineering is a distinct profession, it is important to realize that engineering is not a single activity. Rather it is a wide range of activities. At one extreme the engineer works side by side with the pure scientist—the physicist, the chemist, the mathematician. At the other extreme the engineer rubs shoulders with the technologist—the draftsman, the programmer, the stress analyst. The engineering spectrum includes the research engineer working on the frontier of knowledge as well as the operations engineer responsible for the fine tuning of a multimegawatt power plant.

The classification of engineering into *branches* of aeronautical, chemical, civil, electrical, industrial, mechanical, mining, and petroleum engineering is in terms of what the engineer works *with*. From the standpoint of the student a much more meaningful subdivision is by *functions,* which are related to what the engineer *does*. The major engineering functions are as follows:

Research
Development
Design
Construction
Production
Operation
Application and sales
Management

Because this classification is of such great importance to the student contemplating a career in engineering, these functions will be discussed in detail in Chapter 6.

SUMMARY

The engineering profession can be described and defined in terms of two brief definitions:

• Engineering is the professional art of applying science to the optimum conversion of natural resources to the benefit of mankind.

• An engineer is a person qualified by aptitude, education, and experience to perform engineering functions.

These statements provide the basis for a more detailed study of engineering as a career.

ASSIGNMENTS

1 Write down a dictionary definition of engineering. Explain why this definition is inadequate for our purposes here.

2 Study the definitions and note the date of the first mention of the following words: "science" (or "scientific"), "economy" (or "economic"), and "systems."

3 **a** Study the definitions by Lindsay and Calhoun, and list specific points on which they agree and specific ways in which they differ.

  b Repeat for the definitions of Wellington and Hoover.

  c Repeat for the definitions of Waddell and the Engineering Council for Professional Development.

4 From a recent newspaper or news magazine select a specific technological project. Describe the project in one paragraph; then determine whether it is engineering, as defined here, by discussing it in terms of the six elements of the definition.

5 Interview an engineer (or read a biographical sketch) and obtain specific information about his or her position. Discuss the engineer's work in terms of the definitions in this chapter, answering the questions "Is he or she an engineer?" and "Is he or she practicing engineering?"

6 Vannevar Bush said: "Science never proves anything in an absolute sense." He pointed out that from assemblage of data gained by observation and measurement, the scientist constructs a hypothesis, a formula expressing the relationships he finds. When further observation shows that the hypothesis is faulty, it is replaced by another that seems more nearly correct. Describe a recent scientific advancement, clearly indicating the weakness of the earlier concept and showing how the new hypothesis is different.

7 Cite a specific example in which the engineer must provide:

  a Maximum efficiency for a given cost

  b Maximum performance for a given weight

  c Maximum safety for a given cost

  d Maximum reliability for a given weight

  e Maximum beauty for given performance and cost.

8 Describe the quality of life of an individual in a technologically undeveloped country—for example, an African villager or an Asian peasant. Compare it

with that of a counterpart in the United States, and draw some conclusions about the effect of engineering on the quality of human life.

9 In *The Myth of the Machine*, Lewis Mumford complains about the social organization that technology requires in the construction of great public works. He claims that the benefits of engineering works are canceled out by the restrictions the creations place on human activities, the resulting displacement of humans by machines, and the work's massive destructive potential. Write two paragraphs supporting or refuting his contention, citing specific examples as evidence in your argument.

10 Another definition of profession is "an occupational group with the right to recruit, train, license, and punish practitioners and the authority to set standards, define proper conduct, and expel offenders." Compare this definition with that which is given in the summary. Which of the commonly accepted professions fall under this definition?

Engineering feats of ancient times
The development of civil, mechanical, electrical,
chemical, industrial, electronic, nuclear, computer,
automotive, and aeronautical engineering
The rate of engineering progress

3

A BRIEF HISTORY
OF ENGINEERING

Let us now look at the background of engineering in terms of early engineers and their accomplishments. A brief review of what has gone before will enable us to understand better the relationships among the various branches of the broad profession of engineering. A detailed description of these fields of specialization as they exist today will be given in Chapter 5.

It has been said that the history of civilization is the history of engineering. Certainly it is true that the highly developed civilizations have all been noted for their accomplishments in engineering.

From the moment you are awakened by your clock radio until you fall asleep in an orthopedically designed bed, you are literally surrounded by the results of engineering. You are clothed in fabrics that were unknown a generation ago, you eat a variety of foods of excellent quality, you are protected by inexpensive vaccines for diseases that used to afflict millions, you are propelled through the air by the power of 30,000 horses, and you command the talents of famous entertainers with a flick of a switch. How did all this come to pass?

ENGINEERING FEATS OF ANCIENT TIMES

Prehistoric beings over a period of several thousand years adapted to their own use many things that occurred in nature. Fire gave them warmth, protection from nocturnal beasts, and a means for cooking food and for signaling. The beam, adapted from a fallen tree, permitted them to build their caves in the most desirable places and to cross turbulent streams. From vines and strips of hide they made the ropes that are essential in so many human activities. Their first tools were probably the knife, the ax, and the needle, all patterned after natural objects. The use of the lever required a high type of intelligence and resulted in a multiplication of human force even as the spear extended reach. The first person to use the roller was probably the genius of the tribe; in fact this person's contribution to the problem of transport would have merited promotion to chief. It is interesting to note that when a modern engineer has to move a heavy machine in cramped quarters, he or she goes right back to a prehistoric counterpart and employs the simple lever and multiple roller.

The ancient peoples had mastered the construction of what we call public works: masonry structures, bridges, highways, canals, tunnels, irrigation and drainage systems, water supplies, docks, and harbors. Some of these accomplishments are known only from written accounts and recorded legends. Other works, constructed of stone and metal, still exist as monuments to their builders.

Architecture

The word *architect* means "chief builder." The early architect had to be able to plan a structure to meet a social need, to be skilled in the art (technology) of building, be efficient in the use of materials, and be able to direct workers—in other words, he was an engineer. The great pyramids in Egypt (3000 B.C.), King Solomon's temple in Jerusalem (1000 B.C.), the Parthenon in Greece (450 B.C.), and the Colosseum in Rome (A.D. 80) are well-known structures that attest to the imagination and skill of their planners and builders.

Hydraulics

Water is always intimately connected with human life. The irrigation dams and canals that made the valley of the Nile a garden spot (2000 B.C.) are among the earliest engineering achievements. Both Jerusalem and Athens were supplied with water from the distant hills by means of aqueducts. The Romans are famous for their 250 miles of aqueducts which were described in detail (A.D. 79) by Frontinius, a Roman surveyor and "water commissioner." It has been estimated that these aqueducts, remains of which are still standing, delivered into Rome over 300 million gallons of water per day.

Highways

The Romans carried their engineering into the many countries they conquered. Of interest are the remains of Roman roads found in England, some still in service as foundations for later surfaces. In the Roman Empire there were at one time nearly 50,000 miles of improved roads constructed according to principles of paving and drainage that are still followed. The administration of such a far-flung enterprise depended upon good transportation and communication.

Metallurgy

Human use of fire in daily life undoubtedly led to the chance discovery of the possibilities of metallurgy. You can imagine the cook's interest in the strange flexible layer of metal that was found after using low-melting-point copper ore in building a fireplace. Bronze was discovered early because copper- and tin-bearing ores are frequently found together. Bronze saws and drills were used in the construction of the pyramids, and reference to workers in brass and iron is made in the Book of Genesis. Iron, with its higher melting point, was harder to work. Rare iron ore was converted with charcoal to wrought iron and then fashioned hot by the blacksmith. Its strength and rarity made iron valuable; when not used in a prized tool or weapon, it was formed into bars that were used as currency.

ENGINEERS IN WAR AND PEACE

Under Napoleon, the French improved and extended their highway system and organized a highway administration. Their roads were used primarily for military purposes, and the term "military engineer" was used to designate the individual who designed and built roads, bridges, and forts.

The words "engine" and "ingenious" are derived from the same Latin root *in generare,* meaning "to create." The early English verb *engine* meant "to contrive" or "to devise." Thus the engines of war were clever devices such as catapults, floating bridges, and assault towers, and the designer was called the "engine-er" or military engineer. His counterpart was the civil engineer, concerned with applications of the same knowledge and skill to devising structures, streets, water-supply and sewage systems, and other projects of benefit to the civilian population.

Civil Engineering

The ancient builders had successfully constructed many stationary public works. Civil engineering progress consisted of improving materials, increasing knowledge of the behavior of materials under load, and, consequently, designing more efficiently. In addition, progress was being made in developing sources of power. Early men and women depended on their own muscles for power until they domesticated the camel, the ox, and the horse. Waterwheels and windmills had been used over a long period of time but only to a slight extent. Even as late as 1850 only 6 percent of America's industrial power was furnished by machines; 79 percent was supplied by animals, and the remaining 15 percent came from human muscles. Today about 97 percent of our power is supplied by machines.

Halfway around the world from the well-known public works of the early Egyptians are the rice terraces at Banaue in Northern Luzon, Philippines. Driven from the fertile lowlands by belligerent invaders, members of the Ifugao tribe found security in this mountain valley. Since rice culture requires periodic flooding, terraces had to be carved from the steep mountain sides and an elaborate system of carefully graded canals and dams had to be built. Fish that thrive in the muddy ponds constitute a second "crop" from the terraces. (*Pan American Airways*)

THE POWER OF EXPANDING STEAM

The end of the eighteenth century and the beginning of the nineteenth saw the great transformation in the economic life of England now called the industrial revolution. Before that time the economy was primarily agricultural, transportation was poor, and manufacturing was carried on in the laborers' homes. The invention of the spinning jenny (1763), the waterpowered spinning machine (1771), and other mechanical devices gave rise to the factory system and created a need for mechanical power. The application of Watt's steam engine (invented in 1769) to the cotton, coal mining, and iron industries greatly increased the productivity of English workers and placed England at the forefront of manufacturing and commerce. In the United States the corresponding transformation did not take place until after 1850.

Mechanical Engineering

Those civil engineers concerned with machines were called *mechanical* engineers, and as new sciences and skills developed, they became specialists in the new art. The key development was the invention of the steam engine, which made available large quantities of cheap and dependable power. The feature attraction of the centennial exposition in Philadelphia in 1876 was a 1400-horsepower Corliss steam engine. Engineers, manufacturers, and bankers gazed on the giant machine and were inspired by the prospect of developing labor-saving devices of all kinds. The new devices, in turn, required more efficient use of fuels, and more emphasis on factory planning; thus, mechanical engineering was established as a major branch.

ENERGY FROM THE SKY

One of nature's most awesome displays of uncontrolled energy is the lightning stroke. Under the proper conditions, winds cause a separation of electric charge that builds up and up until, overcoming all restraint, the charges are equalized with a tremendous evolution of energy in the form of heat, light, sound, and, occasionally, destructive forces. The possibility of converting this energy to the benefit of humankind has always been attractive. Although the engineer has not yet been able to harness this wild energy, all its desirable properties have been duplicated under controllable conditions.

   The fact that amber (*electra* in Greek) when rubbed will attract light objects has been known for over 2000 years. The properties of the lodestone or magnet have been known and utilized for over 1000 years. It was not until the eighteenth century, however, that progress was made in the understanding and use of electrical energy. Benjamin Franklin's famous kite-flying experiment in 1752 established the similarity between lightning and static electricity obtained by rubbing "electric" materials.

In the next 50 years, European scientists laid the foundation of electrical science with their systematic observations of static electricity. The voltaic cell invented by Volta in 1800 made available a steady flow of current, and by 1831, Oersted, Ampere, Ohm, Faraday, and other brilliant researchers had established the following facts: A flow of electric current produces a magnetic field; a changing magnetic field induces an electric voltage in nearby conductors; and the current flowing in a conductor is proportional to the voltage.[1]

Electrical Engineering

The first practical application of electrical energy was in communication—the telegraph credited to Samuel F. B. Morse (1840). The telegraph was operated from bulky batteries of relatively weak voltaic cells. The Xénobe Gramme dynamo (1872), which converted mechanical energy into electrical energy, was the real forerunner of the great electrical power industry. It was quickly applied to lighting, and in 1879 the first central power station was installed in San Francisco to supply Charles Brush's arc lights. These lightninglike devices had limited application, however, and it was Thomas Edison's carbon filament lamp, patented in 1880, that provided the great stimulus for devlopments in generation, transmission, distribution, and utilization of electrical energy. Rapid progress in the art and science of application of electricity produced a new specialist—the electrical engineer.

OUT OF THE TEST TUBE

The chemist has always been a partner of the engineer. Knowledge of the composition of substances, their properties, and methods of producing desirable changes in those properties has been an invaluable aid in engineering. The chemist's contributions are obvious in the development of metals, fuels, foods, protective coatings, and the host of other materials employed by engineers. This work was an important part of the industrial revolution, and the term "industrial chemist" was applied to this background associate.

---

[1] In Franklin's time, electrical experiments were performed in courts of royalty and before large public audiences. In an early biophysics experiment, Franklin discharged Leyden jars through chickens, killing a number of them, and concluded, "I conceit that fowls killed in this manner eat uncommonly tender."

Chemical Engineering

By 1880, the use of chemicals in manufacturing and processing had created the new chemical industry in which the production and processing of chemicals by the ton and thousands of tons was the primary concern. The design and operation of chemical plants were carried out by industrial chemists who had learned some engineering or by mechanical engineers who knew some chemistry. The first education program in chemical engineering was established in 1888 at the Massachusetts Institute of Technology.

At the turn of the century German scientists led the world with their discoveries in the chemistry of explosives, dyes, fuels, and synthetic materials. At the beginning of World War I, American industry found itself unprepared to supply the chemicals needed by the armed forces and by the industries manufacturing ordnance materials. The threat of scarcity resulted in an unbelievably rapid expansion of the American chemical industry and firmly established the importance of the chemical engineer. Today this field has expanded to include all types of plastics, synthetic materials, and ceramics, many of which are oil-based products.

INTERNATIONAL COMPETITION AND PRODUCTIVITY

The industrial growth of the western world following World War II encouraged closer international cooperation as well as competition for world markets. Goods from around the world were freely available in the American marketplace and the customer could choose the best product at the lowest price. No longer did local industry have an advantage, as modern transportation and free trade permitted foreign companies to compete equally with them.

Industrial Engineering

This increased competition spurred the development of a new professional, the industrial engineer, whose role was to combine workers, machines, and materials efficiently to increase the productivity of industry. The industrial engineer does not specialize in one aspect of engineering, like the chemical engineer, but is concerned with the management of industry and the complete production process from the mine to the marketplace. The computer has permitted large-scale modeling to aid the analysis and synthesis of complex systems. The increased need for higher productivity has made this field of engineering particularly important to the future competitiveness of American industry.

ELECTRONS UNDER CONTROL

At the same time (1875) Alexander Graham Bell and his assistant Thomas A. Watson were working on the telephone in Boston, James Clerk Maxwell of Cambridge University announced and demonstrated theoretically that electromagnetic waves travel through space with the speed of light. Maxwell's equations are noteworthy because they were not based on experimental evidence; instead, they were deduced by analogy from the laws of light waves.

The German physicist Heinrich Hertz corroborated Maxwell's theory in 1888 by an ingenious experiment; near an oscillating circuit he placed an almost-closed circle of wire and observed sparking across the gap. In 1895 Guglielmo Marconi connected a Hertz oscillator between earth (the *ground*) and an elevated wire (the *aerial*), and was able to detect waves at a distance

The world's first studio for broadcasting was built by Dr. Charles Herrold in San Jose, California, in 1909. This station, now KCBS, provided news and music to an audience as far as 900 miles away. To broadcast music, the disc jockey in the foreground turned his telephone-type microphone toward the phonograph. The phonograph employed an all-mechanical sound system in which the needle transmitted the recorded sound to a metal diaphragm at the throat of the acoustic horn. The transmitted power was coupled to the antenna through the conical transformer in the foreground, and the antenna leads went through the window. (*Foothill Electronics Museum*)

of over a mile. By 1901 he had successfully transmitted signals across the Atlantic Ocean, and radio telegraphy was launched.

Electronic Engineering

In 1883 Edison had noted a curious fact: A small current of electricity would flow between the incandescent filament and another, insulated, electrode placed within the same evacuated bulb. It was not until the electron had been discovered that the Edison effect received any attention. Shortly thereafter the Fleming valve was used as a sensitive detector of radio waves. The introduction of a third electrode, the control grid, into the vacuum tube is credited to Lee deForest (1906).

DeForest's audion could generate, amplify, and detect audio signals or radio waves and was the basis for the rapid development of communications as a subbranch of electrical engineering. Imaginative inventors created the new world of electronics, employing the versatile vacuum tube in countless ways. Then, in 1948, John Bardeen, Walter Brattain, and William Shockley succeeded in controlling the motion of electrons in a solid crystal of silicon. Five years later they had solved the practical production problems and the tiny transistor began to revolutionize electronics.

THE SUN'S ENERGY

For millions of years the sun has been warming the earth's crust and making possible the life upon it; it has been the primary source of all energy on the earth. The tremendous output of the sun cannot possibly be due to the cooling of a hot body. Only recently have scientists fathomed the mystery of its power.

In 1905, at the age of 26, Albert Einstein proposed his theory of relativity, which leads to the conclusion that $E = mc^2$, or mass (m) and energy (E) are interchangeable and are related by the square of the velocity of light (c). Twenty-seven years later, British physicists John Cockcroft and E. T. S. Walton demonstrated the accuracy of Einstein's theory by bombarding a lithium target with high-energy hydrogen ions and measuring the resultant energy and mass. A minute amount of mass had disappeared, and an equivalent amount of energy had appeared. The sun's energy is believed to be created by a series of similar transformations.

History of Atomic Power

"Atom" means indivisible, and for a long time the atom was regarded as the ultimate particle. Our present knowledge indicates that an atom consists of a central, massive, positively charged nucleus and planetary, negatively charged electrons. The nucleus is made up of neutrons (no charge), protons (positive charge), and a number of other elementary particles. An understanding of nuclear forces and reactions has been the goal of many of the world's leading physicists. Some important dates are as follows:

1932   John Cockcroft and E. T. S. Walton confirmed Einstein's theory of energy creation.

1932   James Chadwick (English) discovered the neutron as a nuclear particle.

1934   Enrico Fermi (Italian) bombarded uranium with neutrons and apparently created a new, heavier, radioactive element (93).

1938   Otto Hahn and F. Strassman (Germans) found that one of the products of uranium bombardment is barium—the atom had been split!

1938   Lise Meitner and Otto Frisch (German refugees in Copenhagen) explained Hahn and Strassman's work as nuclear fission.

1939   Enrico Fermi pointed out the chain-reaction possibility if the fission of uranium by neutrons produces more neutrons.

1941   The United States authorized an all-out effort to develop atomic weapons.

1945   The first successful atomic explosion took place at Alamogordo, New Mexico, after four years of work and an expenditure of two billion dollars.

1952   Norris Bradbury and Edward Teller (Americans) used the heat of atomic fission to fuse hydrogen isotopes into helium.

1954   The first thermonuclear explosion contrived by humans was achieved at Bikini Island.

1956   The first full-scale atomic power plant was inaugurated at Calder Hall, England.

1978   Rapidly escalating crude oil prices focused attention on nuclear power as an alternative energy source.

1979   An accident at the Three Mile Island nuclear power plant caused a widespread controversy over the safety of nuclear power.

Nuclear Engineering

The story behind this brief listing is a fascinating illustration of scientific research aided by engineering know-how. It represents the cooperation of brilliant minds from all parts of the western world. Their work has developed a new source of energy of great scientific, social, and political significance. Whether it ultimately results in a significant benefit to human society remains to be seen, but in a single decade atomic fission moved from the laboratory to the industrial world, and nuclear engineering is now well established.

THE ANALYTICAL ENGINE

There is a continuous thread of development running from the ancient abacus to the modern digital computer. Long known in Asia, the abacus was introduced into Europe about 1000 years ago. It is a counting frame in which pebbles or *calculi* (the source of our word *calculate*) are slid along wires to perform number combinations. A skilled abacus operator can stay ahead of

Rapid improvements in computer speed and versatility continue to be accompanied by dramatic reductions in size and cost. Computers with large-scale integrated circuits are becoming ever smaller, faster, less expensive, and more reliable. Nine years' progress in the development of computer memories is illustrated here. Single chips capable of storing over 64,000 bits of information (64K Random Access Memory) are assembled on circuit boards to provide storage for one-half million words (512 kilobytes) in a modern microcomputer. (*Hewlett-Packard*)

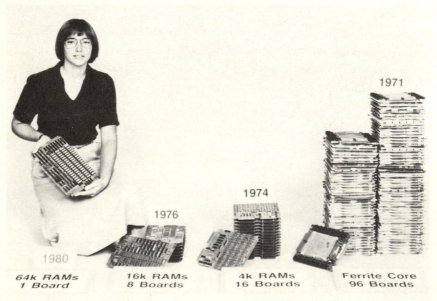

1971

1974

1976

1980

| 64k RAMs | 16k RAMs | 4k RAMs | Ferrite Core |
| 1 Board | 8 Boards | 16 Boards | 96 Boards |

a hand operated calculator in addition and multiplication, and the abacus is still used in the Far East.

The first big step toward the computer was taken in 1642 by a 19-year-old Frenchman, Blaise Pascal, a mathematical genius with an inventive mind. To help his father, a customs officer, in his work, he devised an adding machine consisting of ten numbered wheels linked by gears—the first simple digital calculating machine.

Gottfried von Leibnitz, at the age of 25, invented the stepped wheel (1671) and expanded on Pascal's machine. (A few years later he invented the infinitesimal calculus independently of Newton.) Although this German mathematician's machine attracted much attention, its operation was not reliable. Two hundred years later the Swedish engineer Odhner invented his pinwheel mechanism and developed the fast and dependable hand calculator still widely used throughout the world.

Automatic Calculation

It is difficult to comprehend how much time went into the production of the first tables of logarithms, trigonometric functions, and astronomical data. Many arithmeticians devoted their lives to this repetitious work. A great forward step was breaking down the calculations into basic operations and reducing the work to addition and subtraction that could be done by unlearned persons. In 1812 Charles Babbage, an English mathematician then 20 years old, conceived a mechanical Difference Engine to perform routine calculations and to check the values in published tables (notorious for their errors). Although Babbage's designs were practicable, the technology of his day did not permit the construction of the mechanisms he devised and his machine was never completed.

Babbage's interests were far-reaching. He invented the speedometer, the cowcatcher, and operations research. He advocated mass production, the use of interchangeable parts, and the flat-rate postal system. However, his creativity is best measured by the invention in 1833 of his Analytical Engine, the first universal digital computer. In thousands of detailed designs he described a machine that would accept data on punched cards and, by combining the processes of arithmetic and logic, perform any conceivable calculation. If Babbage's ideas had been understood and appreciated by his contemporaries and if precision tools of the type he advocated had been available, the automatic digital computer would have been possible one hundred years earlier.

Babbage got the idea to use punched cards for recording numerical data from the Jacquard automatic loom. To lift the warp threads to form a pattern, Jacquard used punched cards to push against a bundle of rods and raise those that did not pass through the holes. This idea was rediscovered by Dr. Herman Hollerith of the Census Bureau and applied to tabulating and reducing data. Data from the 1890 census were punched into cards, which were sorted and analyzed by special machines devised by Hollerith. The census analysis was completed in one tenth the time that the previous method would have required and this success stimulated Hollerith to look for other applications of the punched card technique. A company he formed to manufacture accounting machines evolved into IBM.

Computer Engineering

The abacus, Babbage's engine, and Hollerith's punched cards are *digital* devices in that data are represented by discrete units. The slide rule and Babbage's speedometer are *analog* devices in that a length and a dial reading are related by continuous functions to a logarithm and a speed, respectively. The earliest analog computer was the Differential Analyzer invented by Vannevar Bush in 1925. This mechanical computer could solve the sets of differential equations representing complex physical systems.

The machine that actually realized Babbage's dream of an automatic digital computer was the Harvard Mark I conceived by Dr. Howard Aiken and designed and built by IBM. In this electromechanical computer, put into service in 1944, the operations were controlled by switches and relays. It performed 200 additions per minute, worked to 23 significant figures, and could calculate a sine or cosine in about a minute.

ENIAC, the Electronic Numerical Integrator and Computer designed and built by Professor John Mauchley and graduate student Presper Eckert at the University of Pennsylvania, was the first all-electronic computer. It was built in wartime secrecy to calculate urgently needed data on ballistic trajectories. Completed in 1945, ENIAC employed 18,000 vacuum tubes and 6000 switches to perform 5000 additions per second. Although the machine performed up to expectations, it had taken $2^1/_2$ years to build and it was technically obsolete at the time of completion. Subsequent computers were faster, more versatile, more dependable, smaller, and less expensive. In the past decade alone, the speed of computation has increased by a factor of a million and the cost per calculation has decreased by a similar factor! Nearly every business and institution now uses data processing equipment and many families now have home computers.

WHEELS, WINGS, AND THRUST

King Solomon, Julius Caesar, Charlemagne, Queen Victoria, and President McKinley had one thing in common: They all depended on the horse-drawn wheeled vehicle that, in various forms, served humanity for nearly 5000 years. Leonardo da Vinci sketched tanks and flying machines (1480) and Jules Verne described submarines and spaceships (1865), but the realization of these fanciful dreams has been a slow step-by-step process of engineering.

Railroad Engineering

Watt's successful engine spurred application of steam power to water, land, and air transportation. The engineering problems in adapting the steam engine to shipboard use were the least difficult, and by 1786 several inventors had constructed mechanical boats. Robert Fulton was an interested observer at several demonstrations, and after careful analysis of the difficulties came up with the winning combination of a dependable Watt engine, an improved hull design, and strong political backing. From its first run on the Hudson River, August 17, 1807, Fulton's *Clermont* was a financial success.[2]

Steam-powered vehicles had been devised for land operation, but the weight and size limitations were severe. With the advent of high pressure engines and iron rails, the steam locomotive became practical. Robert Stephenson studied the models of earlier builders and then investigated locomotive traction. He discovered that even a 1 percent grade tripled the tractive force required. Keeping grades low, he and his father built several successful English railroads. With their improved locomotive, the *Rocket,* incorporating a fire-tube boiler, advanced firebox design, and forced draft, they won a £500 prize competition in 1829 and set the pattern for all future locomotives.

Automotive Engineering

The personal vehicle that put America on wheels resulted from a series of interrelated developments. Nikolaus Otto perfected his classic *Silent* four-stroke flame-ignited internal-combustion engine in Germany in 1876. The popularity of the bicycle had brought tubular frames, wire-spoked wheels, chain drive, pneumatic tires, and better roads. In 1885 the German engineer Karl Benz built what is generally accepted as the first reliable automobile, a three-wheeler with a single-cylinder four-stroke engine incorporating water cooling, electric ignition, and a differential gear. How does this list of new features compare with those proclaimed for the current model of your favorite make?

American inventors also worked on horseless buggies, with Charles and Frank Duryea, Henry Ford, and Hiram Maxim among the first to make significant contributions. The automobile was particularly suited to America's ge-

[2]Richard Kirby et al., *Engineering in History*, McGraw-Hill, New York, 1956.

ography and independent spirit, and Ford's assembly-line methods put the Model T within popular reach. By 1960 there was a car for every family (on the average) and a network of streets, roads, highways, and freeways beyond the wildest dreams of the most ambitious Roman.

Aeronautical Engineering

The Montgolfier brothers' hot-air balloon (1783), Giffard's steam-powered dirigible (1852), and Zeppelin's rigid airship (1900) were important milestones in our long struggle to overcome gravity, but the faster, more manueverable airplane gave man wings. Here again a combination of related developments set the stage for success. Otto Lilienthal devoted 20 years to careful study and daring experimentation with gliders and lost his life when his craft fell out of control. Octave Chanute, an American bridge engineer, performed gliding experiments that resulted in the biplane design with curved airfoils. Professor Samuel Langley experimented with large model airplanes, some powered by small steam engines.

Wilbur and Orville Wright took advantage of the aerodynamic studies of Lilienthal, Chanute, and Langley and performed their own systematic wind-tunnel experiments and glider tests. They built their version of Otto's lightweight internal-combustion engine and arranged bicycle chains and sprockets to drive two counter-rotating propellers. By warping the wings of their biplane, the Wright Brothers were able to control its flight far better than previous experimenters. On December 17, 1903, the first heavier-than-air flying machine carried man 100 feet across the dunes and into a new era of transportation.

Astronautical Engineering

In his science fiction novel, *From the Earth to the Moon,* Jules Verne in 1865 described rockets for steering a spaceship. Carrying its own oxidizer as well as fuel, the rocket can operate outside our atmosphere. Some of the basic rocket research and tests were carried on by Dr. Robert Goddard from 1914 to 1940 in conjunction with his studies of the upper atmosphere. Amateur rocketry developed rapidly in Germany after World War I (development of conventional weapons being banned by the Versailles Treaty), and with this background Werner von Braun was instrumental in the development of the V2 buzz bomb. After World War II, rocket engineering advanced rapidly and by January 1956, the United States announced plans to launch a series of satellites for scientific purposes. On October 5, 1957, newspapers around the world reported the following announcement by Moscow radio: "On 4th October the first satellite was successfully launched by the USSR . . . . At the present moment the satellite is looping the earth along an elliptical trajectory, . . . making one complete revolution in 1 hour and 35 minutes . . . " On April 12, 1961, Major Yuri Gagarin rode the 10,419-pound Vostok I into orbit, around the earth at 18,000 miles per hour, and safely back. On May 25, 1961, President John F. Kennedy said in a speech announcing Project Apollo: "Now is the time to take longer strides, time for a great new American enterprise, time for this nation to take a clearly leading role in space achievements, which in many ways may hold the key to our future on earth." On January 20, 1969, American Neil Armstrong set foot on the moon, a major step towards the exploration of the universe. By 1980 the Russians had established a fully operational space station. In 1981 the space shuttle provided America with a means of economically developing space opportunities. The tools, vehicles, and knowledge to explore space are available. The door to space exploration had been opened. Whether the American people will step through that door, or let others do it, remains to be seen but astronautical engineering is a reality.

THE RATE OF ENGINEERING PROGRESS

Primitive beings used just what was available, whereas the modern engineer starts with a need and develops a means of satisfying that need. In general, efforts at progress can be classified under four successive levels of activity—utilization, adaptation, conversion, and creation. Engineering, as we have defined it with emphasis on *applying science,* is usually required in the last two stages.

For example, the fallen tree was utilized as a bridge and later adapted to house construction by setting a timber beam on two posts. In the simple truss, the timber was converted to a new form in order to take better advantage of its inherent properties. The alloy-steel cantilever bridge makes use of a material and a form that do not exist in nature but were created for the purpose.

Similarly, we can imagine prehistoric man attracted by the tantalizing odor of a deer roasted in a forest fire and utilizing the result. Adapting fire to a convenient and controllable form of heat energy and making it a part of daily living followed. Conversion of heat energy to the more useful mechanical form in the steam engine was a great accomplishment. At the highest level we could place the creation on earth of temperatures and pressures comparable to those in the central regions of the sun to cause two hydrogen atoms to fuse into one atom of helium, with loss of mass and release of enormous amounts of energy.

In this brief history of engineering only the highlights have been described. Today we are so completely surrounded by the results of engineering achievement it is difficult to realize that we are not at the zenith. Actually we are at a low point on a curve that is rising ever more steeply. The underlying pattern of development reveals that the rate of scientific progress and engineering achievement is ever increasing and probably there will be more new developments in the next 50 years than in the 50 centuries since the construction of the pyramids. And the great contributions are going to be made by young people who, like yourselves, are just beginning a career in engineering. What will be your contribution?

REFERENCES

1 Armytage, W. H. G.: *A Social History of Engineering,* M.I.T., Cambridge, Mass., 1961.
2 Butterfield, Herbert, *The Origins of Modern Science,* Free Press, 1965.
3 Cardwell, D. S. L., *Turning Points in Western Technology,* Science History Publications, New York, 1972.
4 DeCamp, L. S., *The Ancient Engineers,* Ballantine, New York, 1980.
5 Eiseley, Loren, *Darwin's Century: Evolution and the Men Who Discovered It,* Doubleday, Garden City, New York, 1958.
6 Finch, J. K., *Engineering and Western Civilization,* McGraw-Hill, New York, 1951.
7 Gest, A. P., *Engineering: Our Debt to Greece and Rome,* Cooper Square, New York, 1963.
8 Gossick, B. R., "Benjamin Franklin, Scientist," *IEEE Trans. Educ.,* June—September, 1964.
9 Lloyd, G. E. R., *Early Greek Science: Thanks to Aristotle,* Norton, New York, 1974.
10 Lloyd, G. E. R., *Greek Science After Aristotle,* Norton, New York, 1975.
11 Mason, S. F., *A History of the Sciences,* Collier, New York, 1966.
12 Poirier, René, *The Fifteen Wonders of the World,* Margaret Crosland (trans.), Random House, New York, 1961.
13 Rifkin, Jeremy, *Entropy: A New World View,* Bantam, New York, 1981.
14 Rosenberg, Natham, *Technology and American Economic Growth,* M. E. Sharp, White Plains, New York, 1972.
15 Struik, D. J., *Yankee Science in the Making,* Collier, New York, 1962.

ASSIGNMENTS

1 Technological advances are important insofar as they affect individuals or society. Choose an outstanding engineering achievement of the past decade and describe, in a paragraph, the events leading to it. In a second paragraph, discuss the probable effects on human beings.
2 Pick a recent technological invention and draw its "family tree," indicating the discoveries and inventions that led to it.
3 A study of history should reveal trends that, when extrapolated into the future, will aid in predicting what is going to happen. Select an important engineered product (airplane, telephone, computer, refinery, space vehicle, heart-lung machine) and describe the changes in its characteristics that have taken place over a stated period of years. Predict its characteristics at some specified time in the future.

4 Somewhat like our modern astronauts, the Pilgrims ventured into an unknown and hostile environment, and half of them died during the first winter. Assume that you suddenly find yourself (with your knowledge of modern technology) in Plymouth in 1620. You wish to aid the colonists by developing something useful. Select a possible project and, limiting yourself to what was available at that time, outline a step-by-step program for developing the necessary materials and components. Estimate the time required for completion of your project.

5 Prepare a chart showing changes in the speed of travel by businessmen over the past century. Label each period with the most probable mode of travel at the time. Extrapolate into the future, if possible.

6 Helmer and Gordon of the RAND Corporation have forecast the development, over the next 40 years, of the following: directed-energy weapons (lasers, particle beams), commercial ocean-bottom mining, synthetic protein food production, regional weather control, legislating through an automated plebiscite, newspaper printing in the home. Discuss one of these future developments pointing out why it is desired or needed, what recent advances have made it possible, and what further achievements will be required to make it possible.

7 "A radical change in the evolution of civilization followed the invention of the printing press, and we can expect to see a similar radical change in the course of civilization as a result of the invention of the computer." Discuss this statement, pointing out similarities between the two inventions and citing examples of change induced by each.

8 The development of civilization has closely paralleled human use of energy and the discovery of new forms of energy. Estimate the date at which humans began to use fire, water, wind, steam, internal combustion, and nuclear power. Assume that each of these represents an equal upward step in sophistication and plot on a logarithmic scale of time measured in years ago. (Start with one year ago at the right-hand end of your time axis and extend it to 100,000 years ago on the left.) Draw a conclusion about new energy forms expected during your professional career.

9 a Compare the time spans between the first statement of the basic principles and its commercial application for each of the following: steam turbine, photograph, nuclear energy, transistor, digital computer, rocket.

b Draw a conclusion about the time span between concept and application of devices in the future.

Technical societies
Legal registration
Professional ethics
Professional obligations

4

ENGINEERING AS
A PROFESSION

Approximately one million American engineers exhibit an enormous variety of professional specialties ranging from designing electronic circuits to building dams, from devising theoretical models for systems analysis to testing new plastics, from conceiving new means of utilizing solar energy to selling machine parts. . . . There are teachers and deans, brilliant teams in "think tanks," advisors to presidents, titans of industry, rugged individualists heading their own consulting firms; and there are thousands of frustrated inspectors for government agencies, and checkers of quality control in factories. Yet I visualize this vast, motley group as being part of one great profession. . . .[1]

Although the engineering profession is proud of its past achievements, of greater importance is what engineers are accomplishing today. Of even more significance to beginning students is what contributions they and their associates will be making in the years to come. The next five chapters are planned to provide a background for deciding where in engineering your interests may lie. First, we shall look at the whole engineering profession in

[1] Samuel C. Florman, *The Existential Pleasures of Engineering,* St. Martin's, New York, 1976, p. 148.

terms of technical organization, legal registration, and professional ethics. Then, in Chapter 5, the various branches of engineering will be discussed in detail. In Chapter 6, engineering will be described in terms of functions, with comments on the different qualifications and training required for work in each. The relationship between engineers and their close associates—the technicians, technologists, and the scientists—will be explained in Chapter 8 which discusses the technical team. In Chapter 9 we shall provide a statistical picture of engineering including its growth, characteristics, and compensations.

TECHNICAL SOCIETIES IN ENGINEERING

We have seen how civilization has been influenced by engineering achievements. The present form of the engineering profession reflects the same pattern of development. The rapid development of nuclear fission is an example of the way in which scientists, even in different countries, cooperate in bringing about progress. Two men in Germany repeated experiments made in Italy. Their new discovery was interpreted by scientists in Denmark and brought to a meeting of physicists in America. This interchange of ideas among workers in a given field is essential. The modern organization that provides for the sharing of ideas is called a *technical society.* A field is not said to be firmly established until a medium of this kind exists.

Formation of Engineering Societies

In 1902, the fifty-year-old American Society of Civil Engineers heard the president at its annual meeting eulogize the engineer for making human life "not only longer, but richer and better worth living . . . and in the future, even more than in the present, will secrets of power be in his keeping, and more and more will he be a leader and benefactor of men. . . ."[2]

The first technical society in engineering was the Institute of Civil Engineers established in Great Britain in 1828. Its founding reflected the new importance of the civil engineer and may be said to mark the formal beginning of the engineering profession. Engineering is a relatively new profession in comparison with the ministry, law, medicine, or architecture.

The age of engineering in the United States dates from the founding of the American Society of Civil Engineers (ASCE) in 1852. This was a period of great activity in developing waterpower, surveying new regions, and building railroads and canals. Mining is one of the oldest engineering activities, and it was an important segment of the ASCE until 1871, when, with both civil and mining fields expanding so rapidly, the separate American Institute of Mining Engineers was established. This later became the American Insti-

[2]Ibid., p. 3.

tute of Mining and Metallurgical Engineers, reflecting the increased stature of metallurgy. Still later its scope was broadened and its name changed to the American Institute of Mining, Metallurgical, and Petroleum Engineers (AIME) in response to the increasing importance of fossil fuels and energy resources.

The industrial revolution created a need for specialists in power and machinery, and the professional requirements of these workers resulted in the establishment in 1880 of the American Society of Mechanical Engineers (ASME). For many years, the ASME has had an industrial engineering section and various other organizations have had industrial divisions; however, in 1948 the American Institute of Industrial Engineers (AIIE) was founded to serve the particular interests of this branch of engineering. The ASME also had an aviation section, but many aeronautical engineers became members of the Institute of Aeronautical Sciences (IAS), founded in 1932. In 1963, the successor to the IAS merged with the American Rocket Society to form the American Institute of Aeronautics and Astronautics (AIAA) to serve all members of the aerospace community.

The rapid development of electricity as an important energy source and the need for interchange of knowledge in the new technology brought about the American Institute of Electrical Engineers (AIEE) in 1884. The subsequent rush to develop the communications capabilities of electricity led to the establishment of the Institute of Radio Engineers (IRE) in 1912, just six years after deForest patented the three-electrode vacuum tube. In 1962, the AIEE and IRE merged into the Institute of Electrical and Electronic Engineers (IEEE) in order to serve their members better.

The American Chemical Society was founded in 1876. However, the importance of chemistry in manufacturing and processing brought new problems, new methods, and new products that called for the organization of chemical engineers into a tighter group. As a result, the American Institute of Chemical Engineers (AIChE) was organized in 1908.

Until recently, all work in nucleonics was under the control of the Atomic Energy Commission of the United States government. It directed the activities of hundreds of chemical, mechanical, and electrical engineers and performed many of the functions of a technical society. With the shift in emphasis from military to civilian applications of nuclear energy, a new institute was needed to serve the special interests of nuclear engineers and the American Nuclear Society was founded in 1955.

Since the technical society is based on serving special interests, a great many organizations have been established as special subfields. Literally hundreds of specialized societies (engineering, science, management, professional, and technical), organizations, and social groups exist that are indirectly related to engineering. Some organizations, including most of the major engineering societies, have individual student and professional memberships. Many technical societies also provide for industrial, institutional, and organizational memberships and some associations are limited to organizational memberships that represent their constituencies through elected or appointed officials. Table 4-1 lists some of the major engineering societies, including the dates founded, the number of members in 1980, and the types of membership available. A more complete list of science, engineering, and engineering related organizations is given in Appendix A. A list of engineering college honor societies is given in Appendix B.

Functions of Technical Societies

Professional societies should take a more active role in fulfilling the engineer's moral obligation to the public. However, the various professional societies have not been lacking in expression of civic concern. But since the professional societies represent large numbers of engineers, and since there is often violent disagreement among the members concerning what is "good," the societies are subject to severe limitations on their actions. So the role of the professional societies is seen to be most complex and not susceptible to simple solutions in terms of improved social responsibility. The engineering approach cannot solve all social problems, however, it makes an extremely valuable contribution to public discourse.[3]

How do the technical societies serve the engineering profession in general and their members in particular? The fundamental service is in providing for the exchange of ideas and information among workers in a specialized field. The one invention that probably contributed most to the advancement of science and engineering was the printing press. When information could be recorded and widely distributed, it was no longer necessary for each person to start out fresh in the search for knowledge; the individual could "stand on the shoulders" of those who had worked to develop the technology.

Each society publishes at least one journal, an account of the work of its membership. Articles in the journal may be written solely for publication, or they may be papers presented at meetings and selected subsequently for publication. The second major activity is the series of national, regional, and local meetings carried on by each society. These meetings feature announcements of recent discoveries, presentation of new and perhaps controversial ideas, exhibits of novel devices or processes, visits to engineering projects and plants, and reports of committees assigned to study certain problems. In other words, the meetings are for the purpose of keeping engineers informed of the new developments taking place in their special field (Table 4-2, page 49).

[3] Ibid., p. 28.

TABLE 4-1
LIST OF MAJOR ENGINEERING SOCIETIES AND ORGANIZATIONS

Abbreviation	Name of society and address	Date began	No. of individual members/ type of membership
AAES	American Association of Engineering Societies, 345 East 47th Street, New York, NY 10017	1980	1,000,000*
ABET	Accrediting Board for Engineering and Technology (formerly ECPD), 345 East 47th Street, New York, NY 10017	1932	O*
AIAA	American Institute of Aeronautics and Astronautics, 1290 Avenue of the Americas, New York, NY 10014	1932	26,400 S, O†
AIChE	American Institute of Chemical Engineers, 345 East 47 Street, New York, NY 10017	1908	47,400 S
AIIE	American Institute of Industrial Engineers, Inc., 25 Technology Park, Atlanta, Norcross, GA 30092	1948	24,400 S
AIME	American Institute of Mining, Metallurgical and Petroleum Engineering, 345 East 47th Street, New York, NY 10017	1871	73,600 S
ANS	American Nuclear Society, 555 North Kensington Avenue, La-Grange Park, IL 60525	1954	12,000 S, O
ASAE	American Society of Agricultural Engineers, 2950 Niles Road, St. Joseph, MI 49085	1907	9,000 S

*This organization does not provide individual membership.
†S = Student; O = Organizational membership available.
Source: Directory of Engineering Societies, American Association of Engineering Societies, March 1982

In addition, the technical societies are concerned with the professional status of engineers and their education. One important activity is establishing and supporting *student chapters* wherein the engineering student interested in a certain branch is given an opportunity to become acquainted with engineers and their work. The student member receives the journal and

Abbreviation	Name of society and address	Date began	No. of individual members/ type of membership
ASCE	American Society of Civil Engineers, 345 East 47th Street, New York, NY 10017	1852	78,000
ASEE	American Society for Engineering Education, 11 DuPont Circle, Suite 200, Washington, D.C. 20036	1893	10,800 S, O
ASME	The American Society of Mechanical Engineers, 345 East 47th Street, New York, NY 10017	1880	85,000 S
IEEE	The Institute of Electrical and Electronics Engineers, 345 East 47th Street, New York, NY 10017	1884	184,000 S
NACME	National Action Council for Minorities In Engineering, Inc., 3 West 35th Street, New York, NY 10001	1980	O*
NAE	National Academy of Engineering, 2101 Constitution Avenue N.W., Washington, D.C. 20418	1964	1100
NCEE	National Council of Engineering Examiners, P.O. Box 1686, Clemson, SC 29631	1920	O*
NSPE	National Society of Professional Engineers, 2029 K Street, N.W., Washington, D.C. 20006	1934	77,000 S
SWE	Society of Women Engineers, 345 East 47th Street, New York, NY 10017	1950	3300 S, O

is invited to attend the various meetings and field trips. The submission of student papers is encouraged and rewarded by recognition and, occasionally, prizes. Some societies publish student journals directed toward the interests of its student members. Participation in the chapter activities is of great benefit to students because their interest is stimulated, their classroom study is motivated, they learn early what work in the field involves, and they make acquaintances who will help them get started in the work of their choice.

Qualifications For Membership

The qualifications required vary a great deal from society to society; however, there are broad similarities. In general there are four grades of individual membership: Student, Associate, Member, and Fellow. Student membership ordinarily requires attendance at an accredited school of engineering. The rank of Associate is frequently an initial step for those who graduated with a bachelor or higher degree from an engineering institution. The Member grade requires having been "in responsible charge of important engineering work" for a given length of time (3 to 6 years) and being qualified to "design and direct" such work. Certification as to these qualifications is given by society members who are personally acquainted with the applicants and their attainments. Engineering societies confer the rank of Fellow upon a limited number of workers in the field whose accomplishments are outstanding and widely recognized. Honorary membership is also conferred by some societies for unusual and distinguished contributions to engineering.

Unity Of The Engineering Profession

Much of the credit for engineering progress must go to the technical societies for their efforts in stimulating original work, providing for its presentation to interested audiences, and making it available in permanent form. However, the specialization of interest into numerous subbranches has been at the expense of the engineering profession as a whole. There exists no organization for the 1.4 million engineers similar to the American Medical Association (AMA) for physicians, the National Education Association (NEA) for teachers, and the American Bar Association for lawyers. Yet most surveys of engineers have indicated they would prefer to have an "umbrella" organization in addition to the specialized organizations such as ASCE, IEEE, and ASME.

There have been several attempts to unify the engineering profession. In 1920, the five major, or Founder, societies (ASCE, ASME, AIChE, IEEE, and AIME) formed the Federation of American Engineering Societies and set up the American Engineering Council. The AEC operated in Washington, D.C., and was making progress in representing the engineering profession in legislative matters until the depression resulted in loss of its financial support; it was finally abandoned in 1938.

In 1934 the National Society of Professional Engineers (NSPE) was founded. Full membership is limited to those recognized by registration as

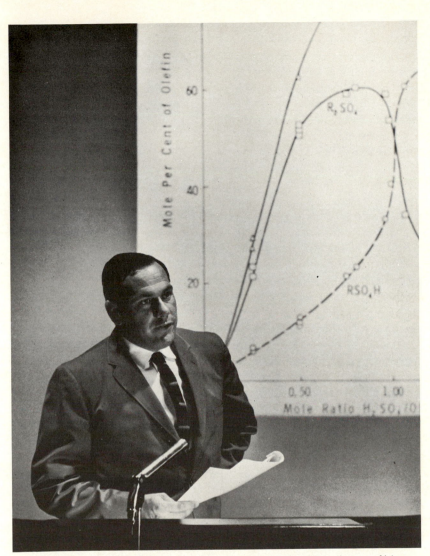

The primary function of a technical society is to provide for the exchange of ideas and information among its members. Here a chemical engineer presents a paper describing his experiments with an improved process for obtaining olefin, an essential ingredient of high-octane gasoline and a fundamental building block of the new plastics. His results are summarized in the form of graphs projected from slides; the data from a series of experiments are identified by the differently shaped points used in plotting the curves. After his presentation, he will be questioned by society members who are interested in specific aspects or by members with conflicting data. Subsequently, an edited version of the paper will appear in the journal of the society. The larger technical societies publish a monthly journal for papers of general interest and a series of specialized publications to meet the needs of the various subgroups. (*Standard Oil Company of California*)

professional engineers in the various states. With headquarters in Washington, D.C., the NSPE has been very active in representing the engineering profession, performing many of the services for which the American Engineering Council was established.

The Engineer's Council for Professional Development (ECPD) was organized in 1932 to promote the professional growth of potential engineers from precollege counseling through college, to initial employment, and finally, to full status as members of the profession. ECPD included the five Founder societies, the American Institute of Aeronautics and Astronautics, the American Institute of Industrial Engineers, the American Society for Engineering Education, the National Council of Engineering Examiners, and the Engineering Institute of Canada, and other engineering related organizations concerned with engineering education and professional development. One of its major functions was accrediting engineering curricula that met its standards of content and quality. In 1980, ECPD was reorganized as The Accrediting Board for Engineering and Technology (ABET), a part of the new American Association of Engineering Societies (AAES).

Another movement, initiated by the Founder societies in 1941, resulted in the creation of the Engineers' Joint Council (EJC). The purpose of the EJC was to "correlate the professional aims of the [Founder] societies and jointly to act for them on matters of national import affecting professional engineers." EJC made various efforts to broaden its base to include other engineering organizations and to provide additional sources to the various engineering societies and their members. One of the most important of these was the organization of the Engineering Manpower Commission which conducts a wide variety of studies on the engineering profession and work force including engineering and engineering technology salary surveys and annual surveys of engineering college enrollments and degrees. In 1980, the American Association of Engineering Societies (AAES) was formed and EJC dissolved. The new AAES has been reorganized into a group of professional, educational, international, and public relations councils that promise new vistas for the professional development of engineers in the United States. Other professional societies are expected to join AAES in the establishment of a single organization that represents and speaks for the overall engineering profession.

The NSPE and the now defunct EJC worked together to secure in July 1952 the congressional recognition of engineering as a profession. This resulted in professional engineers being placed on the same basis as members of the medical and legal professions in regard to wage and salary controls.

In 1953, independent collective bargaining groups of engineering employees organized the Engineers and Scientists of America, a federation of engineers' unions, but this was dissolved in 1961 following the formation of the Engineers and Scientists Guild, a similar organization. Although various efforts have been made to organize engineers into independent and affiliated collective-bargaining groups or unions, the number and proportion of engineers belonging to such organizations has been relatively small with no appreciable growth since the early 1960s.

Concern over public confusion about science and engineering and recognition of the increasing role of engineering in government decision making led to the creation of the National Academy of Engineering (NAE) in 1964. Along with the National Academy of Sciences (NAS), a private nonprofit organization chartered by Congress in 1863, the NAE advises the federal government, upon request, in all areas of science and engineering. A major objective of the NAE is to identify the changing needs of society, to assess the technical resources available, and to sponsor programs of research and development to meet these needs. The primary qualifications for election to NAE membership are important, past contributions to engineering theory and practice, significant leadership experience in engineering organizations in industry, government, and education, or unusual accomplishments in pioneering new fields of technology. Starting from a founding membership of 25, the academy had over 400 distinguished engineers in 1981.

It is now clear that one characteristic of strong but specialized technical societies is their tendency to produce a segmented engineering profession. As a result, a major problem facing engineers is how to preserve the advantages of the technical societies and still bring about a unified and integrated profession that can adequately serve engineering and society.

REGISTRATION OF ENGINEERS

In some European countries, membership in the professional society carries with it the license to practice in that particular branch. In the United States, however, the power to license professional practitioners is vested in the states, and each state has its own law providing for the registration of professional engineers. The first registration law was enacted in Wyoming in 1907 as a result of the chaotic condition that developed when homesteaders surveyed their water rights and signed their own names as "engineers." In 1947 Montana became the forty-eighth state to provide for registration. At the time of their admission, Alaska and Hawaii already had registration acts as do the District of Columbia, Guam, Puerto Rico, and the Virgin Islands.

The purpose in registration or licensing is to protect the public by ensuring that those professing a knowledge of engineering science and a skill in the art of applying this knowledge are competent to practice. The requirements for registration usually include being of good moral character, having satisfactorily practiced engineering in a responsible capacity for a stipulated length of time, and passing an examination of technical qualifications. In general, the examinations are designed to reveal if a person is qualified to plan, design, and supervise the construction or installation of engineering projects to provide safety, economy, and efficiency. Those registered under the law may use the title Professional Engineer (PE).

Less than one-third of those eligible are now registered engineers, in contrast to the nearly 90 percent registration of lawyers and doctors (Table 4-2). The reason is that one does not have to be registered to practice engineering. Subordinates of a registered engineer are not required to register; a large engineering firm may comply with the law by having only its chief engineer registered. Officers and employees of the United States may practice engineering in connection with government service without being registered. Persons serving the engineering profession rather than the public by performing research need not be registered. Some states permit the practice of engineering as long as the person does not call himself an engineer. However, there has been a strong movement in recent years to encourage more engineers to register. Professional registration of engineers has survived every state legislative review. The sunset review hearings of the late 1970s, in which states examined whether or not many existing laws and regulations were necessary, have all resulted in continuation of engineering and land surveying registration.[4]

There are definite advantages in becoming registered. Many public positions demand registration as a qualification for appointment. Some companies have adopted policies specifying registration as a requirement for promotion to senior engineering positions. Courts have held that engineers may not testify as expert witnesses unless they are registered. A contract involving payment of fees for engineering has no legal standing unless the engineer is registered. The Taft-Harley Law gives professional engineers certain rights with regard to collective bargaining. In most states, engineering plans cannot be accepted for official filing unless the signer is a registered engineer. One interpretation by a federal agency stipulates that *competent* engineer and *registered* or *professional* engineer are synonymous and the first qualification is registration. It appears that the trend is toward making registration the accepted criterion for judging the professional engineer.

Engineering registration can be an important factor in an engineer's career development. In the United States, most of the professional engineering examinations are coordinated by the National Council of Engineering Examiners (NCEE). Professional engineering registration in most states consists of passing two examinations: (1) the Fundamentals of Engineering which is based largely on an engineer's academic training, and (2) the Principles and Practice Examination which is based on the engineer's early professional ex-

[4]Donald A. Buzzell, "Trends in Engineering. Registration Sunset Review. How to Survive?" *Amer. Soc. Civil Eng.*, Preprint 80-508, New York, NY, 1980.

TABLE 4-2
PERCENTAGE OF ENGINEERS REGISTERED, HOLDING
MEMBERSHIP IN PROFESSIONAL SOCIETIES, AND PARTICIPATING
IN ON-THE-JOB TRAINING AND CONTINUING EDUCATION BY
BRANCH OF ENGINEERING (1972)

Branch	Professional societies	Registered	Continuing education
Aeronautical	38	18	63
Agricultural	77	38	47
Chemical	62	24	49
Civil	56	64	57
Electrical	44	22	61
Industrial	37	20	64
Mechanical	41	33	56
Metallurgical/Materials	71	19	49
Mining & Petroleum	77	44	58
Nuclear	67	23	55
Other	44	28	54
All branches	47	32	56

Source: The 1972 Scientist and Engineer Population Redefined, vol. 1, Demographic, Educational, and Professional Characteristics, National Science Foundation, NSF 75-313, U.S. Government Printing Office, Washington, D.C., 1972.

perience. The NCEE invites engineering seniors to take the Fundamentals of Engineering exam just prior to graduation—or as soon thereafter as possible. It is no secret that the best time to take the Fundamentals of Engineering exam is during the closing months of an undergraduate engineering program when problem-solving techniques for a variety of engineering subjects are fresh in mind. Seniors who take and pass the Fundamentals Exam (leading to certification as Engineers-in-Training) may apply to take the Principles and Practice Examination 4 years later. The time to start the registration process is in your undergraduate years. Don't put it off until a job opportunity arises which requires registration and then say, "Why didn't someone tell me?"

Fifty-four states and/or territories of the U.S. have enacted individual legislation and most will grant reciprocity and/or comity if there is a change in the state of employment or residence.[5]

As a result of recent legal implications relating to product liability, and to the qualifications needed to serve as expert witnesses, an increasing number of engineers will find registration essential in order to maximize their career options.

[5] National Council of Engineering Examiners, Engineering Registration and the Fundamentals Examination, Seneca, NC, 1980.

PROFESSIONAL ETHICS

There was, for a time, resistance among some engineers to promulgation of formal ethical codes, it being said that the professional gentleman's conscience was the best guide in such matters.[6]

Responsible engineers have long recognized the necessity for a statement of principles of professional practice that would serve as a guide in personal conduct. In 1906, the American Institute of Electrical Engineers appointed a committee consisting of President Schuyler S. Wheeler, H. W. Buck, and Charles P. Steinmetz to draw up a code of ethics. Similar groups representing other engineering societies have worked over the years to develop standards to meet the needs of their members.

In 1963, the Engineers' Council for Professional Development (ECPD) adopted a detailed code of ethics that spelled out in considerable detail the obligations of engineers to the public and to each other. In 1977, after considerable discussion and debate, a shorter code (see page 51) was approved by the board of directors of ECPD, and it has since been adopted by most of the engineering societies. ECPD, now ABET, also adopted and has modified a set of guidelines for use with the Fundamental Code of Ethics.[7] In 1979 the National Council of Engineering Examiners developed a set of *Model Rules for Professional Conduct* for registered engineers (see Appendix 4C). The mutual obligations between employers and engineers are spelled out in *Guidelines to Professional Development of Engineers and Scientists.*[8]

PROFESSIONAL OBLIGATIONS

Of particular interest to you as a beginning engineer is the obligation of experienced engineers to make available their knowledge to others. Only a minute part of your ultimate understanding will come as a result of your own original work; by far the greatest portion will be received from others—in school, through engineering societies, and from other engineers. In turn, you will be obligated to make your contribution to the constantly expanding store of knowledge. Engineers also have an obligation to remain up-to-date by participating in continuing professional development. In recent years, this has become increasingly important due to rapid developments in science, engineering, and technology. Engineering societies, colleges and universities, employers, and other specialized technical and management organizations have developed a wide range of self-study materials, short courses, workshops, media packages, and other programs and services

---

[6]Florman, p. 19.

[7]Engineers Council for Professional Development, "Suggested Guidelines for Use with the Fundamental Code of Ethics," 47th Annual Report of ECPD, 1979, pp. 26–27.

[8]National Society of Professional Engineers, Washington, D.C., 1980.

CODE OF ETHICS OF ENGINEERS

THE FUNDAMENTAL PRINCIPLES

Engineers uphold and advance the integrity, honor and dignity of the engineering profession by:

I. using their knowledge and skill for the enhancement of human welfare;

II. being honest and impartial, and serving with fidelity the public, their employers and clients;

III. striving to increase the competence and prestige of the engineering profession; and

IV. supporting the professional and technical societies of their disciplines.

THE FUNDAMENTAL CANONS

1. Engineers shall hold paramount the safety, health and welfare of the public in the performance of their professional duties.

2. Engineers shall perform services only in the areas of their competence.

3. Engineers shall issue public statements only in an objective and truthful manner.

4. Engineers shall act in professional matters for each employer or client as faithful agents or trustees, and shall avoid conflicts of interest.

5. Engineers shall build their professional reputation on the merit of their services and shall not compete unfairly with others.

6. Engineers shall act in such a manner as to uphold and enhance the honor, integrity and dignity of the profession.

7. Engineers shall continue their professional development throughout their careers and shall provide opportunities for the professional development of those engineers under their supervision.

Approved by the Board of Directors, October 5, 1977

designed to help engineers advance professionally, and to guard against potential technical obsolescence. As may be seen in Table 4-2, engineers have met their professional obligations in varying degrees, with regard to professional society membership, engineering registration, and participation in continuing education.

> If by chance the day should arrive when the basic bodily needs of all men are fulfilled—and how unlikely that prospect appears at this moment—the engineer's opportunity to contribute to the welfare of mankind will still be limitless. There will always be a need to plan for future development and maintenance of resources, to safeguard against creation of new weapons, and to maximize the opportunities for people, not only to survive, but to pursue their interests in the fields of education, art, and recreation.[9]

Engineering is more than an occupation; it is more than a vocation. Engineering is a way of life. As a young engineer you will be indebted to a host of pioneers, both renowned and anonymous, who have made the technical, professional, and ethical advances that have made engineering an honored and respected profession. The fruits of their efforts are available to anyone with the necessary aptitude who will seek them with proper diligence. The only payment expected is a dedication to the continuation, extension, and elevation of the profession of engineering.

REFERENCES

1 Florman, Samuel, C., *The Existential Pleasures of Engineering,* St. Martin's, New York, 1976, p. 148.

2 Buzzell, Donald A., "Trends in Engineering. Registration Sunset Review. How to Survive?," *Amer. Soc. Civil Eng.* Preprint 80–508, New York, New York, 1980.

3 National Council of Engineering Examiners, Engineering Registration and the Fundamentals Examination, Seneca, NC, 1980.

4 National Council of Engineering Examiners, *Model Rules of Professional Conduct: A Guide for Use by Registration Boards,* Seneca, NC, 1979.

5 Alger, L., Christenson, N.H., and Olmsted, P., *Ethical Problems in Engineering.* Wiley, New York, 1965.

6 American Association of Engineering Societies, *Directory of Engineering Societies and Related Organizations,* New York, 1981.

7 National Council of Engineering Examiners, "The Registration of Professional Engineers and Land Surveyors in the United States," 56th Annual Meeting, Atlanta, Ga., 1977.

8 National Council of Engineering Examiners, *Model Law,* Seneca, NC, 1978 (rev.).

9 Engineers Joint Council, *Guidelines To Professional Employment For Engineers And Scientists,* 2d ed., August 1978.

10 National Science Foundation, *The 1972 Scientist and Engineer Population*

[9]Florman, p. 146.

Redefined, vol. 1, *Demographic, Educational, and Professional Characteristics,* NSF 75–313, 1975.

11 Engineers Joint Council, *Directory of Engineering Societies and Related Organizations,* New York, June 1979.

12 Association of College Honor Societies, *Booklet of Information,* Library of Congress Number 76-8325, Williamsport, Pa., 1980–1981.

ASSIGNMENTS

1 Select a technical society in which you might be interested in the future, and answer the following questions:
  a When was the society founded?
  b What technical developments led up to the founding of the society?
  c What contributions has the society made to engineering?

2 Select a technical society whose journal is in your school or public library. Study six recent issues of the journal, and answer the following questions:
  a What types of articles appear in the journal?
  b What topic is of greatest current interest?
  c Generally speaking, what positions are held by the authors of the articles in the journal?

3 Compare the requirements for membership in the American Association of Engineering Societies with one of the other technical societies.

4 Explain why practicing doctors are uniformly licensed whereas less than a third of the qualified engineers are registered. Is this difference in status in the public interest? Explain.

5 Study the legislative act under which professional engineers are licensed in your state. Determine the legal basis for licensing as carried on by your state. Determine the legal basis for licensing as carried on by your state, and summarize the requirements for licensing or registration as a professional engineer.

6 One reason for enacting a legal description of engineering is to prevent encroachments by other professions that are legally recognized. Cite specific instances where, in the absence of legal definitions, there could be conflicts of interest between engineering and architecture, accounting, the medical profession, the legal profession, and real estate appraising.

7 Study the Canons of Ethics of Engineers and answer the following questions.[10]
  a An engineer works for a large company in a newly developing area of technology and is familiar with recently developed devices and processes and other trade secrets. He is offered a position with a small company that would obviously benefit from his knowledge. What are his obligations to his former employer if he changes employment?

---

[10] See L. Alger, N. H. Christenson, and P. Olmsted, *Ethical Problems in Engineering,* Wiley, New York, 1965.

b An engineer whose staff is lightly loaded with work offers to provide professional services on a project of the highway department of another state. She is informed that a specific political contribution is expected. She subsequently learns that this practice is common knowledge among engineers in that state. What should be her course of action?

c In April, an engineering student accepts an offer of employment starting in June with the ABC Co. In May he receives an invitation to visit the very attractive XYZ Co. at the company's expense. What courses of action are open to him?

d An engineer believes that a proposal made by the city engineer of her community is technically unsound. What course of action should she follow?

e An injured workman asks an engineer to appear on his behalf before a workmen's compensation board. The determination rests upon certain technical details related to an accident. The workman states that he cannot afford to pay the engineer for her services. Is it ethical for an engineer to provide services as an expert witness for an indigent client on a contingent basis (whereby she would be paid a percentage of any award)? On a free basis?

8 Some professions have the right to train, license, control, and expel members. How do these prerogatives affect the medical profession? How would such power affect the engineering profession?

Civil   Materials   Mechanical
Industrial   Aerospace
Electrical   Chemical   Agricultural
Architectural   Nuclear   Biomedical   Computer

5

BRANCHES OF ENGINEERING

One justification for a book describing the nature of engineering is that this broad field of knowledge covers so many diverse activities. As indicated in Figure 5-1 one way of classifying engineering is in terms of major and minor branches, each branch being concerned primarily with a particular type of product, process, or industry. In many cases, arbitrary decisions have been made in designating the major branches and subbranches and in choosing where a given specialty will be listed. It must be realized, however, that there is a great deal of overlapping, and few engineers can afford to limit themselves to a single field of specialization.

The branches of civil, materials, mechanical, industrial, aerospace, electrical, and chemical engineering are described in detail, and the distribution of engineers by industry within fields is given in Figure 5-1. The work done in certain subbranches is described briefly.

CIVIL ENGINEERING

Civil engineering occupies a prominent position in the history of engineering, and at the present time nearly one-seventh of all engineers are engaged primarily in this branch.

Civil engineering

Structures	Pipelines	Waterways	Energy
Mechanics	Highway	Construction	Oceanography
Hydraulics	Geodetics	Transportation	Cartography
Soil mechanics	Surveying	Urban planning	Environmental
Geotechnical	Irrigation	Water supply	Sanitation

Materials engineering

Geological	*Metallurgy*	*Materials Science*	
Mining	Beneficiation	Structures:	Properties:
Petroleum	Extraction	Thermodynamics	Mechanical
Ceramics	Processing	Kinetics	Electrical
Plastics	Fabrication	Synthesis	Optical

Mechanical engineering

Air conditioning	Gas turbine	Automotive	Heating and
Fluid mechanics	Nucleonics	Machine design	ventilating
Applied mechanics	Power plant	Steam turbine	Product design
Thermodynamics	Heat transfer	Instrumentation	Machine tools

Industrial engineering

Robotics	Manufacturing	Safety	Methods
Computer-aided	Production control	Time and motion study	Human factors
manufacturing	Product safety	Facilities design	Operations research
Cost control	Quality control	Job evaluation	Plant layout

Aerospace engineering

Astronautics	Structures	Missile and space	Air transport
Celestial mechanics	Power plants	systems	Aerophysics
Aerodynamics	Controls	Environment	Propulsion

Electrical engineering

Automata	Power systems	Feedback controls	Electroacoustics
Bioelectronics	Illumination	Solar devices	Circuit theory
Quantum electronics	Solid state	Computers	Communication
Integrated circuits	Microwaves	Instrumentation	Control systems
Energy conversion	Machinery	Industry applications	Microprocessors

Chemical engineering

Process development	Paints	Petroleum	Food processing
Heavy chemicals	Plastics	Chemical processes	Process control
Process design	Rubber	Pollution control	Biosynthesis

Agricultural | ## Nuclear | ## Computer | ## Architectural

Agricultural	**Nuclear**	**Computer**	**Architectural**
Land use and drainage	Power plant	Information systems	Building design
Food engineering	Nuclear safety	Digital computers	Equipment
Agricultural machinery	Nuclear materials	Computer design	**Biomedical**
Solar and energy	Fusion	Software	Bioengineering
Farm structures	Reactor control	Image processing	Medical
Electrification	Radiation	Computer graphics	Clinical

FIGURE 5-1
Classification by major branches and subbranches of engineering.

The breadth of civil engineering is shown by the wide range of topics included in the tabulation of subbranches. The civil engineer is directly concerned with society and the environment; and civil engineering projects are frequently characterized by extreme size, complexity, durability, and cost.

The wide diversity of activities in which civil engineers are organized is reflected in the specialties of members of the American Society of Civil Engineers which are summarized in Table 5-1 including structures, construction, highways and transportation, geotechnics (soil mechanics), hydraulics and water resources, and urban planning and development.

Structural Design

The largest single group of civil engineers (see Table 5-1) is concerned with the design of commercial, industrial, and public buildings; bridges; dams; towers; tanks; missile testing and launching facilities; etc. These are built of steel, concrete, timber, earth, brick, plastics, glass, and aluminum in a great variety of forms including columns, beams, girders, trusses, arches, membranes, shells, and slabs. The structures must carry their own weight plus that of their intended loads; in addition, they must withstand earthquakes, vibrations, wind pressures, wave impacts, snow loads, blast forces, and destructive radiations.

TABLE 5-1
DISTRIBUTION OF CIVIL ENGINEERS BY SPECIALTY

Specialty	Percent
Structural	18.3
Construction	14.6
Highways and transportation	11.7
Geotechnics, soil mechanics	9.3
Hydraulics and water resources	9.1
Environment and sanitation	7.9
Urban planning and development; city planning	5.6
Engineering mechanics	4.1
Soil mechanics	9.3
Irrigation and drainage	9.1
Waterways, port, coastal, and ocean	3.0
Surveying and mapping	2.8
Energy	2.5
Pipelines	2.2
Other	5.7
Total base number	81,342

Source: Official Register, 1981 American Society of Civil Engineers, New York, 1980, p. 99.

The creations of the structural engineer must be compatible with the anticipated environment: extremes of climate on the earth, tremendous pressures at the bottom of the ocean, near vacuum and absolute zero in space, and unknown conditions on the distant planets. Continual innovation in design is necessary to obtain desired performance, with improved safety, over extended time periods, and under strict cost restraints. Computers have made it feasible to examine a wide range of configurations, materials, and design that make it possible to optimize on a wide range of factors including cost, safety, and time.

Construction

The American construction industry has led the world in the development of equipment for the rapid and efficient completion of projects that are grand in concept, huge in size, and complex in nature. The construction engineer makes use of trucks, conveyors, cranes, and hoists; trenching machines and other earth-moving equipment; machinery for making, placing, and finishing concrete; equipment for welding and riveting; and the special skills of a great variety of craftworkers blended into an efficient team. Now statistics, operations research, computers, and other modern tools of systems planning are being employed to increase efficiency.

Highways and Transportation

Transportation is a key aspect of modern society, and each of its subbranches poses problems requiring economic studies of the population to be served, anticipated traffic, potential hazards, and possible site locations. Highways must be aesthetically pleasing as well as fast and safe. Reducing traffic deaths and injuries on highways and within cities continues to concern and challenge highway engineers. Airport runways must withstand the pounding of heavy cargo jets and supersonic transports. Launching facilities for trips into space, the forerunners of public spaceports, present new challenges to the transportation engineer.

Preliminary surveys are made of topography, soil, weather conditions, and land costs. New methods of surveying and mapping, some using laser beams or satellites, are more accurate and far more rapid. Efficient design requires knowledge of soils, foundations, structures, materials, construction methods, and drainage systems, as well as vehicles and the humans who operate and use them. Great progress is being made in understanding the structural behavior of the soil, which ultimately supports nearly all engineering structures. An old transportation means, the pipeline, is assuming new importance in the efficient movement of coal, cement, grain, and diverse

Erecting an antenna on a polar mount to serve as a ground link in the international satellite communication system illustrates the wide range of civil engineering responsibilities. After creating an effective structure on paper, the engineer must choose materials and shapes to provide strength, rigidity, and weather resistance at minimum cost. Design of the foundation depends on site surveys and a knowledge of soil mechanics. Time and safety become important factors for the engineer supervising construction. (*Pacific Telephone*)

refinery products, as well as water, oil, and gas. Systems analysis, automation, and computers are now employed routinely in solving transportation problems.

Hydraulics and Water Resources

Hydraulics and water resources engineering has to do with the collection, control, utilization, and conservation of water; it includes flood control, irrigation, drainage, reclamation, storage, navigation, recreation, hydroelectric power production, municipal water and sewage systems, pumping machinery, and hydraulic motors. The water resources engineer is called on to predict population growth and its effect on water need, to plan economically feasible water distribution systems, and to design dams and canals that will conserve water and also preserve the landscape.

With the growth of our population and the increase in our standard of living, water is rapidly becoming a critical resource. At present, we use less than 10 percent of the rain that falls on the continental United States, but our needs are increasing every year and in some dry areas we will need to use and reuse every drop of available water—use it for recreation on the lakes in the mountains, for power production as it falls, for water supply and irriga-

A new source of water may result from studies at this pilot plant designed and constructed by sanitary engineers with support from the U.S. Public Health Service. The effluent from a municipal sewage plant provides the input for this 10-gallon per minute plant which has demonstrated that potable water can be reclaimed for around 17 cents per 1000 gallons. The pilot plant permits experimentation at reduced cost and with smaller time constants. Scaling up the design to final operating size will require changes in materials and processes as well as in size. A critical factor in the success of this project will be overcoming public reluctance to drink "water from the sewer." How would you solve this problem? (*Stanford University Photo*)

tion in the valleys, for industrial processes in the cities, and for transportation on the rivers. And after water has been used, contamination must be removed so that it can be used again and again.

Environmental and Sanitation Engineering

As people crowd together in urban communities, they are more exposed to the undesirable side effects of technology. Environmental engineering involves the application of engineering knowledge and techniques to the promotion of good health and the prevention of disease. Sanitation engineers are responsible for the control of the environment in the interest of public health and, therefore, are concerned with water supply, food and milk production and distribution, sewerage, waste disposal, water and air pollution, and the health and safety aspects of housing—plumbing, ventilation, and heating.

The technical interests of environmental engineers cover an extremely broad area, combining engineering talents with a working knowledge of medicine, biology, physics, and chemistry. They are challenged by new problems posed by nuclear and chemical wastes, air and water pollution, and revolutionary changes in food processing; however, environmental and sanitation engineers must solve their problems within the old limitations of cost and public approval. When their problems are solved successfully, they receive the special satisfaction of knowing that people are directly, if unconsciously, benefited.

Urban Planning and Development

Proper design of its elements—roads, buildings, schools, airports, and water treatment plants—will not alone ensure an efficient, safe, pleasant, and beautiful city. Urban planning is the coordination of engineering, architecture, and landscape design in the comprehensive planning of new communities and the redevelopment of existing cities. City planning involves legal enactments and practical politics on one hand and aesthetics and sociological research on the other.

The effects of urban crowding, deficient housing, inadequate transportation, and insufficient open areas on the inhabitants of our great cities are all too clear. Urban development engineers may find their most challenging problems in the renewal of blighted communities. However, solving problems as they arise, by corrective action, is no longer an acceptable approach. Urban planners must think in terms of comprehensive designs for large geographical areas, taking into account diverse functional requirements and the special needs of minority groups, anticipating future demands, and considering society's need for aesthetic surroundings.

International Development

By definition, underdeveloped nations lack the public works that have contributed so much to the welfare of the citizens of more affluent countries. Dams can provide the assured water supply needed for increased food production and for inexpensive power production. Highways decrease transportation costs and increase the commerce that builds economic strength. Modern water and sanitation systems greatly reduce disease incidence and increase the health and effectiveness of humans. Civil engineers are needed in all parts of the world to consult with government officials, to advise engineers and train them in modern methods, and to supervise construction by international firms.

The Future of Civil Engineering

In the past fifty years civil engineers have made great progress along two lines: understanding of the physical principles basic to the behavior of their designs and skill in the technology of their profession. With this knowledge, civil engineers have designed tilting and rotating microwave antennas as large as bridges, pumps that move molten sodium as if it were water, structures to be moored at the bottom of the sea, vehicles to carry payloads into space, and power and water supplies for operation on other planets. Civil engineers are taking advantage of the capabilities of shaped charges in excavation, television cameras in exploratory boring, laser beams in tunneling through rock, operations research in mass transportation planning, and computers in coordinating traffic signals with vehicle flow.

The interactions of technology and society present civil engineers with another series of problems. They must develop more efficient methods of placing concrete, fabricating structural elements, and applying attractive finishes. Civil engineers must learn more about prestressing and continuity, about rainfall and runoff, about the ocean and its resources, about economics and financing, about regional planning and the prevention of blight in larger cities. They must solve pressing problems in mass transportation, waste disposal, water supply, air pollution, and international development. Certainly future civil engineering calls for technically competent women and men with great vision.

MATERIALS ENGINEERING

Our modern civilization requires a continuous stream of basic materials from the earth. As you look around your room, practically everything you see was dug or pumped out of the earth or came from the earth as an agricultural product. Mining, ceramics, and metallurgy are among the oldest engineering arts, but when applied to the resources on the moon or at the bottom of the ocean they are part of one of the most modern branches of engineering.

The organization of the broad field of materials engineering is reflected in the wide diversity that characterizes the professional specialties within the American Institute of Mining, Metallurgical, and Petroleum Engineers (AIME) which is organized into four major constituent societies: the Society of Mining Engineers (35 percent), the Metallurgical Society (14 percent), the Society of Petroleum Engineers (47 percent), and the Association of Iron and Steel Engineers (4 percent). In addition, there are many other related engineering, scientific, and technical societies concerned with materials engineering that are not directly associated with AIME including the National Institute of Ceramic Engineers, the American Society of Metals, the American Association of Petroleum Geologists, and the American Association for Testing and Materials. This diversity not only within engineering but in closely allied fields such as ceramics, geology, metallurgy, and petroleum characterizes the extreme complexity and diversity within materials engineering.

To provide a clear but somewhat oversimplified picture of the activities of engineers in this broad branch, let us apply the term *materials engineering* to the general category and define[1] the subdivisions as follows:

Geological engineering involves the study of rocks and soils to determine the structure of the surface and subsurface. The geological engineer obtains information from soil analyses, excavations, valley walls, wells, aerial photographs, and geophysical measurements using gravitational, electric, magnetic, and seismic instruments.

Mining engineering includes the exploration, location, development, and working of mines for extracting fuels such as coal, lignite, and oil shale; metallic ores such as copper, iron, lead, gold, silver, tin, uranium, and zinc; and other minerals such as asbestos, bauxite, potash, and sulfur.

Petroleum engineering includes exploration, drilling, production, storage, and transportation of crude petroleum and natural gas.

Ceramic engineering includes preparation of nonmetallic minerals from raw materials; forming by presses, molds, and wheels; firing in kilns, ovens, and furnaces; and application to industrial and domestic uses.

Plastics engineering includes the preparation, fabrication, and application of natural and synthetic materials that deform permanently under stress, particularly cellulose derivatives and polymers that are easily molded.

Metallurgical engineering includes production of metals from ores by mechanical, thermal, electrical, and chemical processes; development of

[1] American Society of Mechanical Engineers, "Definitions of Occupational Specialties in Engineering," New York, 1951.

metallic alloys with needed characteristics through knowledge of molecular and crystalline structure; and fabrication of metal products by casting, welding, and powder metallurgy.

Materials science is the general, quantitative, science-based study of materials—their structure, processing, and properties—and the synthesis of materials to meet specified needs.

Geological Engineering

The old-time prospectors with their donkeys, pans, picks, and shovels are nearly gone from the mining picture. They were limited to following clues that appeared on the earth's surface. Modern exploration to find hidden deposits requires scientific training and elaborate equipment such as the seismograph, magnetometer, and Geiger counter, mounted on a truck or helicopter or an earth-circling satellite.

In addition to searching for fuels and minerals, geological engineers apply their knowledge of the earth's structure to engineering problems in the construction of roads, landing strips, skyscrapers, tunnels, dams, and harbors. In these activities they may work for private petroleum and mining companies, for private construction firms, or government agencies, or they may provide service as independent consultants.

Mining Engineering

After the location of an ore deposit is determined, the earth must be "opened up" to discover the nature, extent, and value of the deposit. If the project appears economically feasible, then expensive equipment is moved in to bring the ore to the surface for further treatment. Opening up and removing the ore involves high explosives, power tools, complex machinery, rapid materials handling, elaborate safety precautions, and constant attention to costs. It should be clear that mining engineers depend on the knowledge and techniques of civil, electrical, and mechanical engineers, geologists, and chemists, in addition to those of their own specialized fields. As the need for metals and energy sources increases with the expansion of our economy and as the existing rich deposits are exhausted, they will need help in devising new techniques for locating and processing poorer ores.

Petroleum Engineering

The refining and manufacturing of petroleum and its products will also be considered later as part of chemical engineering, since the great majority of engineers in this field are chemical engineers. Petroleum engineering as used here has to do with exploration, drilling, production, storage, and transportation of crude petroleum and natural gas.

One theory of the origin of petroleum is that it is part of the remains of multitudes of microscopic sea animals. It is trapped under folds (anticlines) or domes of nonporous rock layers. Usually the oil-bearing sand lies above a saltwater area, and above the oil is found a gas pocket. Off-shore drilling involves similar techniques as those of on-shore drilling but is based on contour mappings and sampling of the ocean floor.

Working from maps prepared by geological engineers or oceanographers, petroleum engineers must decide if, where, and how to drill the wells or erect off-shore platforms. The design of drilling equipment is an important responsibility. As the hole is drilled, provision must be made to prevent such occurrences as rapid wear of the bit, caving-in of the walls, intrusion of water, deviation of the size or direction of the bore, and sudden blowouts due to high gas pressure. The great need for skill and judgment is obvious when it is considered that the work of drilling is out of sight—perhaps 5 miles underground. Core samples are taken at suitable intervals, and electrical logging instruments are used during the drilling, but the petroleum engineer cannot go down into the "mine" for a look around.

After the drilling has penetrated into the producing formation to the optimum distance, the well may be "shot" by high explosives to increase production. If natural pressure is insufficient to raise the oil to the surface, suitable pumping means must be selected and installed. Pumping methods, pumping rates, well spacing, and artificial injection of gas, air, or water are carefully controlled to conserve reservoir energy and ensure maximum recovery of oil.

The crude oil is transported to the refinery by pipeline, tank car, tank truck, or marine tanker. Although gasoline is the most important refined product at present, it is recognized that petroleum is a complex raw material that, in the hands of the chemical engineer, may be turned into an endless variety of useful products. Oil is a natural resource without which no modern nation can survive, and its discovery, production, and processing has become increasingly important.

Ceramic Engineering

The term *ceramic* is applied to the products obtained by high-temperature processing of nonmetallic minerals. Included are such diverse items as microcomputer chips, spark plugs, stereo cartridges, glazed tile, dinnerware, grinding wheels, aircraft parts, firebrick, fishing rods, and window glass. One of the oldest engineering fields, it is increasingly important in our industrial and domestic life.

The ceramic engineer is concerned with grading, grinding, and mixing raw materials; forming desired shapes on potters' wheels and in molds and presses; and firing the product at high temperature in kilns and furnaces. The machines and methods for carrying on these processes include applications of various mechanical, electrical, and combustion equipment. The ceramic engineer is responsible for designing and operating the equipment, testing the products, and carrying on research in raw materials, processes, and products.

Abrasives employ natural materials such as corundum and garnet and artificial materials such as silicon carbide and fused alumina. Structural clay products include glazed and unglazed brick, tile, and sewer pipe; the clay is mixed with water, formed, dried, and then fired to develop strength and durability. Whiteware includes glazed or unglazed china and porcelain used for kitchenware, art objects, electrical insulators, and computer memory cores. Refractories are ceramic materials that resist melting and therefore are useful in furnace linings, rocket exhausts, and reentry shields. Glass is an inorganic product that has cooled without crystallizing. Its outstanding characteristics are hardness and transparency, although it may be opaque or colored or in the form of flexible fibers. Cements include Portland cement (used in concrete), asphaltic cement, limes, and plasters. Electroceramics used in transistors and other microelectronics systems are revolutionizing the electronic and computer industries.

Until recently ceramic work was strictly an art with very little science. The ratio of engineers to skilled workers in ceramics was low compared with that in other industries. However, the increased demand for ceramic items and the rapid development of new products have resulted in more emphasis on mass production, and a parallel demand for advanced technology research and development. The introduction of modern machinery has increased the need for technically trained personnel, and the outlook for ceramic engineers is very favorable.

Plastics Engineering

From the invention of celluloid in 1868 and the development of the first synthetic polymer in 1910, the production of plastics in the United States has grown rapidly and now exceeds a billion pounds per year. Polymers are formed by joining two or more molecules of the same kind to produce a new compound having the same chemical composition but different physical properties. These high-molecular-weight materials, called *macromolecules,* consist of long chains of molecules, sometimes branched and sometimes crosslinked, and it is the wide range of resulting properties that makes polymers so valuable.

Plastics engineers are concerned with the formulation, production, and application of polymers. They study the attractive forces developed by valence bonds and the plastic slipping of molecules under stress. They know that increasing the molecular weight of polymers increases the strength and resistance to solvents and decreases the ease of molding and extrusion. They also know that irregular chains are more flexible than regular chains and that products containing flexible chains are soft and rubbery with high resistance to impact. For a given application, the plastics engineer must decide on the optimum compromise.

Plastics are fabricated in a variety of ways. Thermosetting resins are forced into a mold of a desired shape and held under heat and pressure until cured. Thermoplastic compositions are softened by heating and then injected into a cold mold or extruded in the form of films or pipes or fibers. Plastic foams are poured into molds and cast in the desired shape. The final properties depend on the variation of temperature and pressure with time and on subsequent mechanical working and thermal annealing. The tremendous range of properties already available and the potential extension of this range of properties to new extremes make the field of plastics engineering attractive and challenging.

Metallurgical Engineering

In general there are two broad areas of activity: *extractive* metallurgy and *physical* metallurgy. Extractive metallurgists are concerned with the recovery of metals from their ores and their further refining; the output of the mine becomes the raw material for extractive metallurgists. They first use various mechanical means to remove the valuable minerals from the undesirable accompanying material, a process that is called *ore dressing,* or *beneficiation,* meaning "to make better." The location of the beneficiation process and the extent to which it is carried are based on economic considerations. Next, extractive metallurgists must separate the metal from the chemical compounds in which it exists in the ore and obtain commercially pure metal. *Pyrometallurgy* involves the use of fire as a roasting, melting, or reducing agent as in the blast furnace. *Electrometallurgy* is employed in the production of aluminum and in the purification of copper.

Physical metallurgists then take over and put the refined metals to use. They are concerned with making alloys that will yield improved physical properties: high-strength steel, stainless steel, low-melting-point solder, corrosion-resistant bronze, high-permeability iron, high-purity transistor silicon, and easily machined cast iron. Foundry work is a special field of physical metallurgy involving choice of furnaces, metals, molding sands, and core materials. Forging and welding have benefited from the work of the metallurgist in alloying, physical testing, and heat-treating. Heat-treatment is an example of how the engineer can convert given materials to more useful forms. If red-hot steel is allowed to cool slowly, it is usually soft and easily bent; however, if the same steel is cooled quickly or *quenched* in water, it becomes strong and hard. Starting with a given combination of iron, carbon, and minor alloying elements, the metallurgical engineer can, by specifying the rates and degrees of heating and cooling, obtain extreme hardness or ductility. By this art, the range of properties of engineering materials has been extended greatly.

Materials Science

For many years metallurgists have known *which* metals, alloys, and processes would provide strength, conductivity, or corrosion resistance. But *why* is steel strong? *Why* is copper a good conductor? *Why* does stainless steel resist corrosion? While metallurgy is the science of metals, materials science is the science of all materials—metals, ceramics, and plastics included. Answers

This age-hardenable aluminum-magnesium-silver alloy is an achievement of materials science. The samples were heated at 800°F for two hours to permit the components to go into solution. Then the samples were quenched in cold water to provide a soft material that could be easily worked into any desired shape. Aging at 400°F for 16 hours produced this fine dispersion of dark precipitate (visible within the grain boundaries in the 1000× microphotograph on the left) that provides great strength. On the right is a picture of the grain boundaries at 50,000× magnification obtained on an electron microscope. What practical applications can you think of for a material that is easily worked for a while and then gets strong as it ages? (*Reynolds Metal Company*)

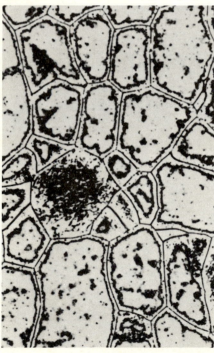 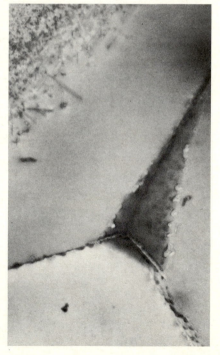

to these questions provide the basis for the understanding of materials and their behavior that will lead to the significant advances in material properties needed for progress in many fields of engineering. This quantitative knowledge of materials, based on physics and chemistry and expressed in terms of general principles, is called *materials science.*

In plastics and ceramics our use is far ahead of our understanding; in the case of crystalline metals, however, materials scientists are beginning to establish the quantitative principles needed to predict the behavior of known alloys and to synthesize useful new ones. They know that the strength of steel depends on the strength of the bonds holding its atoms together, which can be calculated, and the distribution of "dislocations" in the crystalline structure, which can be controlled. They know that the conductivity of copper is directly related to the mobility of the electrons in the outer atomic orbit. In contrast, the conductivity of a p-type semiconductor (silicon alloyed with a trivalent element such as boron) depends on the mobility of an electron vacancy called a *"hole."*

Using this basic knowledge, materials engineers can increase the high-temperature strength of gas turbine blades by specifying that they be made from specially grown single crystals of a nickel-cobalt alloy. They can increase the safety of an aircraft fuselage by specifying rivets made of an aluminum-copper alloy that is soft at first and then becomes harder and stronger with age. They can improve the performance of a Mars-bound satellite by specifying an additive to the solar cells that will make them heal after radiation damage. They can control and put to use the phenomenon of superconductivity that eliminates electrical resistance at temperatures near absolute zero, or the phenomenon of superplasticity that permits certain alloys to elongate without breaking. Because progress in many branches of technology is limited by the availability of substances with the necessary properties, the materials engineer faces a challenging future.

The Future of Materials Engineering

Materials engineering comprises a wide spectrum of activities that are continually changing. The continuing development of materials with outstanding combinations of mechanical, chemical, and electrical properties, provides unusual opportunities for technological advances in almost every engineering frontier. As a result, materials engineers are expected to play key roles and assume major responsibilities in research, development, operations, and management in American industry. The challenging problems in energy, environment, and transportation provide a wide variety of opportunities not only in metal production and fabrication, but also in ceramics, semiconductors, and other nonmetallic materials. The importance of providing new energy sources offers materials engineers national and international opportunities in extraction and refining of petroleum, coal, and other strategic materials. The increasing importance of research and development in materials science also promises to provide excellent future opportunity for materials engineers in microstructural analysis, electroceramics, and chemical metallurgy.

MECHANICAL ENGINEERING

The work of the mechanical engineer is probably more familiar to the general public than that of any other engineer; we are surrounded by illustrations of the mechanical engineer's skill in converting material and energy into desirable things. As may be noted in Table 5-2, mechanical engineers work for a diverse group of employers and have a wide range of occupational titles, responsibilities, and broad as well as specialized interests. No one group dominates; however, in 1981, members of the American Society of Mechanical Engineers were likely to be employed in manufacturing (52 percent), engineering professional or administrative positions with active interest in management (15 percent), design (13 percent), power (10 percent), or applied mechanics (6 percent). From the time of the industrial revolution, the mechanical engineer has been concerned with devising engines that develop power, applying the power to useful purposes by

TABLE 5-2
DISTRIBUTION OF MECHANICAL ENGINEERS BY SPECIALTY

Specialty	Primary, percent	Primary and secondary, percent*
Management	15	10
Design engineering	13	10
Power	10	7
Applied mechanics	6	5
Heat transfer	5	5
Plant engineering and maintenance	5	5
Pressure vessel and piping	4	5
Aerospace	4	3
Fluids engineering	4	4
No primary division recorded	3	1
Nuclear engineering	3	4
Petroleum	3	2
Production engineering	3	4
Gas turbine	3	3
Other	19	31
Total base number	81,859	292,509

*Multiple responses possible.

Sources: American Society of Mechanical Engineers, "ABC Report, Geographical Area," 1981; American Society of Mechanical Engineers, "Inter Office Memo," Technical division counts as of 9/81.

means of machines, using machines in the transportation of persons and things, and employing heat transfer in controlling thermal energy.

Engines

We use the term *engine* to mean a device for developing mechanical work from natural energy sources; this includes transforming the potential energy of falling water in a hydraulic turbine or the fission energy of uranium in a nuclear power plant, as well as converting the chemical energy of a fuel in steam or gasoline engines.

Steam   When a fuel is burned in a boiler or steam generator, chemical energy is converted to thermal energy in the flame, which, in turn, is transferred to water or steam. The energy of the steam is converted to mechanical work in a reciprocating engine or in a rotating turbine. The mechanical energy may be used directly or converted into electrical energy for transmission elsewhere.

Internal combustion   If the fuel is burned within the device itself, it is called an internal-combustion engine. The most important of these are the Otto cycle engine (as in most automobiles), Diesel cycle engine, gas turbine, jet engine, and rocket motor. The rocket motor is unique in that, instead of depending on the atmosphere, it carries its own oxidizer as well as fuel; this permits it to function at high altitudes or in the vacuum of outer space. With demands for energy increasing and known fuel supplies dwindling, increased concern for the environment, mechanical engineers must strive for increased efficiency and low pollution as well as high performance. Their designs work metals near their ultimate strength and use fuels in increasingly efficient processes. They optimize fuel injection and ignition, cylinder-piston-valve combinations, exhaust systems, pumps, cooling-water supply, lubricating systems, and instrumentation.

Machines

Power is valuable when it is used to drive machines that produce desired results. The approach to contriving the machine depends on its intended application. The modern computer has revolutionized the design of machines. Computer design and graphics have increased the ability of the mechanical engineer to obtain complete motions, optimize the use of materials, reduce costs, to improve safety, and to extend the life of modern machines.

Machine Design   In designing a heart-lung machine, the problem is one of invention, creation of a clever system for gently pumping and oxygenating the patient's blood during surgery without damaging the fragile corpuscles. In contrast, the design of an aircraft control system to provide rapid response with dynamic stability must be based upon highly scientific theory and precise data. Design of an automobile body necessitates taking into account passenger safety as well as specific factors of strength and rigidity. Machine designers employ power-transmission elements such as shafts, gears, and torque converters; actuating mechanisms such as valves, levers, and cams; and structural elements such as fasteners, supports, and bearings. They specify standard components to cut cost, or dream up new devices to meet special needs.

Machine Tools   Of great importance are tools that rapidly and precisely fabricate machine parts from raw materials. These include lathes; shapers; grinders; milling, drilling, and boring machines; and the associated cutters, dies, jigs, and fixtures. Of increasing importance are the computer-controlled machines that automatically position work, change tools for a series of operations, and inspect the results. Mechanical engineers have responsibility for the design, production, and application of machine tools and also for the study of cutting and abrasive action, lubricants, and coolants. They must take advantage of new processes such as spark machining, laser-beam drilling, anodic dissolution, and ultrasonic vibration. In addition to rolling, forging, and drawing, metal forming by rapid methods using powerful magnetic and explosive forces is becoming important.

Machinery   The great majority of mechanical engineers are employed in manufacturing. Machines created by mechanical engineers find application in mines miles below the earth's surface and in satellites miles above the surface. Printing presses, linotype machines, textile looms, sawmills, cultivating and harvesting machines, concrete mixers, office machines, and home appliances indicate the tremendous variety of laborsaving devices that have contributed to this country's standard of living. In production, storage, and transportation, equipment is needed to handle items and materials on an occasional or continuous basis. Included are cranes, conveyors, chutes, pipes, pumps, and lifts. The imaginative design and ingenious application of this materials-handling equipment is a special subdivision of mechanical engineering that can yield great economies.

Transportation

Mechanical engineers are concerned with practically every inanimate thing that moves. Of particular importance is the group of machines that move and carry people or things—vehicles of land, sea, air, and space. The railroads and ships formerly dominated transportation, but the number of engineers employed in these industries now is small compared with those in the automotive and aerospace industries.

The 800-mile-long Trans Alaska Pipeline carries oil south across Alaska from Prudhoe Bay on the North Slope to the port of Valdez on Prince William Sound. Recoverable reserves of approximately 9.6 billion barrels are located in the Prudhoe oil field. Careful analysis of eight proposed routes led to selection of the path. In conducting the preliminary studies, and later in the design and construction of the pipeline, engineers had to cope with a harsh climate, rugged terrain, and stringent environmental regulations. More than half of the pipeline is elevated and the remainder is buried below ground. In the early stages of operation, the 48-inch diameter pipeline carried about 750,000 barrels of oil daily. (*Alyeska Pipeline*)

Automotive   First the rich person's luxury, then the working person's necessity, the automobile has become a part of our mode of existence, and the number of two- and three-car families is rapidly increasing. The world energy shortage and environmental concerns have also contributed to accelerating efforts in the engineering design of motor vehicles. Contributing to the production of millions of units per year are multiples of assembly plants, many major manufacturing plants, and hosts of small and large suppliers of component parts. Mechanical engineers are active in design, production, and testing capacities at all stages of manufacture. Because of the intense international and national competition in the automotive industry, large numbers of engineers are engaged in research and development of smog-free engines, more fuel efficient engines, impact-absorbing bodies, smoother transmissions, and better-handling suspensions.

Aerospace   The rapid growth of air transportation over the last three decades can be expected to continue as we strive to shrink the earth and pull the planets into our own sphere. Many individuals with degrees in mechanical engineering are employed by aircraft and spacecraft manufacturers. At present, the largest numbers are in design and development of new aircraft and spacecraft, with smaller numbers engaged in production, operations, and maintenance. Aeronautics and astronautics are dynamic fields where the technology is constantly changing; and engines, aerodynamic structures, and operating controls are subjected to continual experimenting. Probably refinement of design is carried further in aircraft than in any other engineering product because of the great importance of providing high performance, safety, and light weight under conditions of extreme acceleration, speed, and temperature.

Other Transport   A distinguishing characteristic of the mid-twentieth century is the mobility of the earth's inhabitants. We are probing the depths of the oceans, the heights of the atmosphere, and even interplanetary space. A major segment of engineering talent is occupied with devising efficient and safe means for getting people from one place to another. Among the less common means are elevators, escalators, sidewalk conveyors, monorails, chair lifts, and ski tows. Of particular importance are mass transportation systems that hold the key to continued development of metropolitan communities.

Energy

One difference between the poor and the rich is that the rich have their heat in the winter and their ice in the summer. Since the time of the steam engine, the production, transformation, and application of thermal energy to close this gap has been the concern of the mechanical engineer.

Heating and Ventilating   The conversion of fuels to heat is carried on in furnaces and ovens for all types of industrial and domestic purposes. Heat is a valuable ally in metallurgy, ceramics, heat-treating, and refining, as well as a comfortable "friend" in the home. Since the distribution of heat in the form of warm air requires the same types of blowers and ductwork as does the removal of undesirable odors and poisonous fumes, these two fields are usually linked. It is necessary that the heating engineer be skilled in the application of insulation to keep the heat where it is wanted.

Refrigeration   When heat is removed, we have cooling, frequently by refrigeration. The most common domestic refrigerators employ the compression method in which a refrigerant is compressed by mechanical means, cooled in an exterior condenser, and then expanded rapidly, causing it to absorb heat from the food storage compartment. Refrigeration is used in food storage, food preservation, industrial processing, and air conditioning.

Air Conditioning   Human comfort depends on physiological and psychological factors. The mechanical engineer possesses the knowledge and has available the machinery for controlling one part of the human environment. Air conditioning refers to the simultaneous control of air temperature, humidity, movement, and purity. Air-conditioning systems include heaters, refrigerators, washers, filters, blowers, duct work, and controls.

Heat Transfer   The transfer of thermal energy always involves the exchange of heat between two bodies, for example, a gas and a liquid, or a liquid and a pipe wall, or two regions of a turbulent liquid. The design of better heat exchangers depends on a more complete understanding of the complex mechanisms whereby energy is transferred. This fruitful field for mechanical engineering research includes topics such as the mechanics of bubbles, turbulent and laminar flow, ionization of air by high temperatures during atmosphere re-entry, boiling under gravity-free conditions, and heat transfer across an electrically conducting fluid in a magnetic field.

Energy Conversion   Present nuclear power plants use the heat given off in atomic fission to generate steam, which then drives a turbine that turns a generator in the conventional way to produce electrical power. The direct conversion of heat to electricity has long been a dream of engineers, and recent developments have made it possible on a practical basis. *Thermo-electricity* is an old science founded on the observations of Seebeck (1822) and Peltier (1834) that current will flow when the junctions of two dissimilar materials forming a thermocouple loop are maintained at different temperatures. The inverse effect can be used to obtain cooling or heating from the same unit by introducing a flow of current in the proper direction. Among the new semiconductors are materials with excellent thermoelectric properties that will provide reasonable efficiencies after the engineering problems have been solved.

Other promising direct energy conversion devices are under study by mechanical engineers. In the *thermionic* converter, heat supplied to a cathode causes the emission of electrons with sufficient energy to cross a vacuum to the anode and flow through an external load. In the *fuel cell,* the chemical energy of a fuel is converted directly into electrical energy by reaction with an electrolyte in the presence of a catalyst. In the *magnetohydrodynamic (MHD) converter,* heating produces a high-velocity stream of ionized gas, which flows through a transverse magnetic field that drives positive ions toward one electrode and negative ions toward the other. Electrical energy is gained at the expense of mechanical energy abstracted from the moving gas.

The Future of Mechanical Engineering

It appears from the foregoing that mechanical engineering is one of the most varied of engineering fields and possesses almost unlimited potential for development. The major industries are tremendous segments of our national economy that have not yet reached their full growth. Smaller industries show just as much promise. Existing industries will be completely overhauled to take advantage of new developments such as automation, the use of computers, or solar energy. In addition, the advent of new sources of energy, new materials, and new products will create whole new industries undreamed of at the present time and mechanical engineers will play a key role in these developments.

INDUSTRIAL ENGINEERING

The rapid mechanization of manufacturing industries solved many problems of low-cost mass production but simultaneously created many new problems involving the huge capital investments required, the people who operated the machines, the new techniques employed, and complex systems of communication. Such problems constitute the branch of industrial engineering, which is much broader than the name implies. Industrial engineering is concerned with the development, design, installation, and operation of integrated systems of people, machinery, and information in the production of either goods or services.

The employment and interests of members of the American Institute of Industrial Engineers reflect the emphasis and concerns of industrial engineering. Table 5-3 provides some insight into the primary job activities of AIIE members based on a recent survey. The same survey indicated that 14 percent were employed in manufacturing which included that of fabricated metals (5 percent), primary metals (10 percent), and electrical machinery (7

Shhhhh. . . . . Jet aircraft of the 1980s may be markedly quieter as a result of new ideas now being developed at this research and development center. In this versatile acoustic test facility, an engineer prepares to test a model of the turbofan of an advanced jet engine. On the other hand, consumer studies have shown that making some devices, like vacuum cleaners, too quiet has an adverse effect on their sales! (*General Electric Company*)

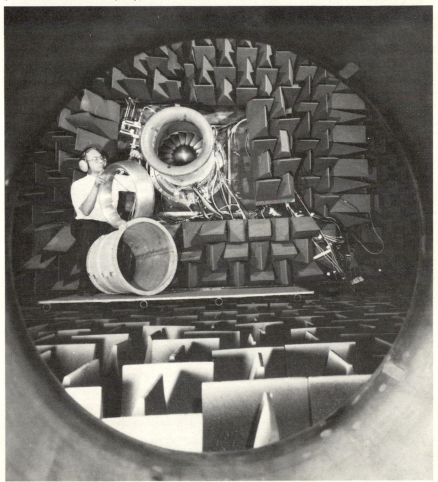

TABLE 5-3
DISTRIBUTION OF INDUSTRIAL ENGINEERS BY
JOB ACTIVITY

Job activity	Percent
Cost control and reduction programs	63*
Work measurement, standards, performance measures	54
Methods engineering, process design, value analysis, human productivity	53
Facility planning, design, layout, environment protection	48
Organization, administration	40
Engineering economics	34
Materials handling	33
Management information systems	29
Manufacturing engineering, automation	25
Incentive plans	21
Safety and health	19
Warehousing, shipping, receiving	19
Employee relations, job evaluation, collective bargaining	19
Production and inventory control scheduling, forecasting	19

*Multiple responses possible.
Source: American Institute of Industrial Engineers, "Reader Profile," August 1981.

percent). About 7 percent were employed in transportation, communication and related services; 6 percent in health, education, and other personal services; and 3 percent in government and public administration. Over 13 percent of the AIIE members were in top management positions; 12 percent were in senior industrial engineering jobs; 31 percent were in industrial engineering title positions; and 13 percent held production manager positions.

Methods engineering is an attempt to apply the same scientific approach to operations involving humans that has been applied so successfully to machines. Motion studies are made to minimize effort and eliminate eliminate energy- or time-wasting movements. Individual work places are designed so that the proper tools are readily available, materials are accessible, and workers are comfortable and safe. The industrial engineer advises in the design of jigs and fixtures to ensure that human factors are considered. Materials-handling methods and whole plant layout become the result of scientific analysis rather than tradition or whim.

No modern plant can stay in business without cost control. From the time the raw materials enter the plant until the finished product is on its way to the customer, a careful accounting must be kept of all incident costs. Industrial engineers must understand cost accounting so that they can interpret excessive costs in terms of faulty plant operation. The trend in modern practice is toward standard costs, which are estimates of what costs should be under ideal conditions. They provide a measure of accomplishment and point the way to lower production costs. Detailed statistics and cost records are important tools of the industrial engineer.

When a part has been produced it must be inspected to see if it meets the specifications established by the engineering department in response to the customer's desires and the manufacturer's capabilities. Quality control is a modern manufacturing concept that goes beyond mere inspection on the premise that making unacceptable parts can, in most cases, be prevented. There is always variability in any physical process and it is impossible to make the objects "exactly" identical. However, in a properly operating process, the variability follows a pattern defined by the average or mean value and the spread or dispersion of values about the mean. Using statistical principles, the industrial engineer establishes acceptable limits for the variation of small samples of parts and devises convenient charts for continuously observing the quality of production. Effective use of quality control reduces spoilage, reworking of defective parts, and customer complaints.

Human Resources

It is the emphasis on human factors that most clearly distinguishes industrial engineering from other branches. At first a major problem was getting humans to adjust to the requirements of laborsaving devices; now the industrial engineer influences the design of machines to harmonize with the capabilities and limitations of the operators. An important area of research is concerned with the psychological and physiological characteristics of the human and machine system. In designing a weaving machine or an automobile—and particularly a space vehicle—the limiting factors in its operation may be the capabilities of the human who controls it.

One method of increasing human efficiency is to make sure that every worker knows what he or she is supposed to do and how to do it. Additional training is always required when a plant shifts over to a new operation or a new procedure. A new method that shows great promise may be a complete failure if the operators are not properly oriented. Management, as represented by supervisors, may need training in carrying out supervisory responsibilities. Industrial engineers may have to educate labor representatives with regard to the merits of a proposed plan.

What performance can be expected of an average qualified worker in a given operation? Time studies may be used by the industrial engineer as a basis for determining standards of work performance. These may be used in conjunction with motion studies in comparing current methods with proposed new methods.

How should the pay of a plant electrician compare with that of a tool-and diemaker, assuming that each does a fair day's work? Since efficiency in production depends so much upon the attitudes of the workers, the rates of pay must be accepted as fair. One responsibility of the industrial engineer may be to develop a basis for comparing jobs in terms of responsibility, training required, and physical hardship. When appropriate scales have been developed, each job can be evaluated and that evaluation interpreted in terms of pay.

Manufacturing, Automation, and Computers

Initially, industrial engineers worked with manufacturing equipment— machines that substituted for the hands, fingers, and backs of human beings. Humans then performed primarily the control functions of starting, stopping, and guiding. As it became possible to design more controls into the machines, they became automatic, and the human operators exercised higher levels of control such as comparing output against prescribed standards, detecting the effect of wear in control mechanisms, and making necessary fine adjustments. In the fully automated machine with feedback control, such comparisons and adjustments are provided by the machine itself and the human operator is freed for a still higher level of activity. With each change in the role played by humans, new problems are presented to the industrial engineer.

Some early factories were portions of a home to which the owner invited neighbors to bring their tools to work on the materials provided. The products were then sold and the proceeds divided among the participants. In such a situation, financial problems were simple. In contrast, the modern manufacturers who compete in the open market must organize teams of skilled workers, make huge investments in machinery, establish continuous sources of raw materials, set up elaborate sales organizations, and make long-range plans. The financial decisions must take into account previous expenditures, current expenses, estimated costs, hoped-for profits, and certain taxes on the profits, if any. The relocation of a plant, the purchase of a new piece of laborsaving machinery, or the introduction of a new product constitutes an engineering problem involving complex economic as well as technical factors. The science of logically considering complex economic factors is called *engineering economics,* and it is one of the basic tools of industrial engineering.

Cars come to life in this framing fixture called a Robogate, which automatically welds together the two sides and underbody panel of each car body. The Robogate and other automatic welding systems apply a total of nearly 3000 welds to each car body to insure high standards of quality. Automated assembly lines like this are required for the car industry to survive and prosper in the 1980s. The assembly lines will require fewer workers who will be highly skilled and better paid. (*Chrysler Corporation*)

Information

Recently, the engineering and scientific communities have been sensitized to the seriousness of lagging U.S. industrial productivity. This problem is evidenced by a weakened competitive posture, increasing world trade deficits, and a stagnating standard of living. As a result of growing attention by the news media, low productivity is becoming a concern of society at large. Its symptoms within the industrial sector are tightly restricted capital investment and manufacturing expansion, vastly reduced R&D expenditures, decreased innovation, and poor quality assurance.

Clearly the problem must be addressed in the most meaningful terms, that is, by upgrading the tools and methods of production. There is general agreement that to do this, and to compete successfully in world markets, American industry must adopt the sophisticated techniques of computer-aided design (CAD), computer-aided manufacturing (CAM), and computer-integrated manufacturing systems (CIMS) on a widespread basis. Robotics is also providing new hands and locomotion to the solution of modern productivity problems. However, the rapidly changing state of the art, the variety and incompatibility of computing hardware, and the enormous cost of software development pose serious obstacles to widespread adoption. Industrial engineers are expected to play a key role in the development of both software and hardware in the future.

As an enterprise becomes larger and more complex, the importance of communication increases. Keeping communication efficiency high and costs low is usually the responsibility of industrial engineers. They are concerned with information on quantities (output, hours worked, inventory), qualities (product, raw material, employee), costs (materials, labor, packaging), instructions (orders, working drawings, policies), and reports (from plants, from sales, to management). They are responsible for devising procedures and equipment for processing all types of information—obtaining it in the field, distributing it throughout the plant, using it in planning, or storing it for future use.

Data processing is a basic activity of industrial engineers. The increasing capacity and availability of computers and other microelectronic data processing equipment influence the work of an industrial engineer in two ways. First, in many companies the industrial engineer is asked to design data processing systems, to select and specify the appropriate equipment, and to write the necessary instructions. Second, computers are valuable aids in solving production problems, in controlling sophisticated machine tools, in keeping accurate records, in forecasting market demands, and in optimizing manufacturing schedules. A major application of computers is in simulating the behavior of complex organizations or production systems so that the effect of adding a warehouse, of automating an assembly line, or of introducing an industrial process can be predicted and evaluated before the changes are actually made.

The Future of Industrial Engineering

The breadth of industrial engineering is evidenced by the wide range of such activities as research in biotechnology, development of new concepts of information processing, design of automated factories, and operation of incentive wage plans. Already, the techniques of industrial engineering have been applied in the construction and transportation industries, in farm management and crop harvesting, in restaurant and hotel operation, in airline operation and maintenance, and in major government activities such as the post office and military supply operations. More and more the particular talents of the industrial engineer will find application in mail-order merchandising and retail store operation, in hospital planning and surgical procedures, in public utilities and banks, and in an ever-widening segment of modern civilization.

AEROSPACE ENGINEERING

Aerospace engineering is another important branch of engineering. A number of different names are used to identify academic curriculums for this professional field. *Aerospace* usually includes studies in the atmosphere (winged vehicles) and in space (rockets). *Astronautics* normally refers to space flight. *Aeronautics* and *aeronautical engineering* normally refer to atmospheric flight.

Just as the crude flying machine successfully flown in 1903 by the Wright brothers gave humanity its long-dreamed-of wings, so *Sputnik I* on October 4, 1957, pointed the way for ever-adventurous humans to overcome the bonds of gravity leading to Neil Armstrong's landing on the moon on July 20, 1969. Today, astronautics and aerospace have become fascinating, challenging, and demanding fields of activity for thousands of engineers and scientists who are launching space laboratories and satellites in near-earth orbits and to distant planets and galaxies.

The employment and specialty interests of the members of the American Institute of Aeronautics and Astronautics (AIAA) provide insight into the employment and subbranches within aerospace engineering. About 80 percent of the 25,000 AIAA members in 1981 were engineers; over one-half (54 percent) were employed in manufacturing including 18 percent in aircraft and 18 percent in spacecraft and missiles; about one-fourth (25 percent) were employed in government (primarily in federal agencies including NASA and civilian and military positions in the Department of Defense); and about one-fifth in universities (8 percent) and other commercial and nonprofit organizations engaged in research and related government contracts. Over one-third of the AIAA members were in administration or management; 25 percent were in technical management, 8 percent were in general management, and 2 percent were in business management. As may be noted in Table 5-4, AIAA members have very wide primary and secondary specialty interests.

Aeronautics

Wilbur and Orville Wright carried on a careful and systematic program of glider testing, control development, and engine building over a period of three years before their first successful flight of less than a minute at Kitty

TABLE 5-4
DISTRIBUTION OF AERONAUTICS AND ASTRONAUTICS
ENGINEERS BY PROFESSIONAL INTEREST

Area of interest	Primary interest, percent	Any interest, percent
Atmosphere and space scientists	14	13
Aircraft design	12	8
Missile and space systems	12	12
Mechanics and control of flight	10	10
Propulsion	10	9
Materials, structures, and dynamics	9	10
Management, history, and economics	8	7
Information systems	7	9
Vehicle support, test, and dependability	6	8
Aircraft systems	4	5
Aircraft operations	3	3
Energy systems	3	3
Application of aerospace technology	1	2
Hydronautics	1	2
Total base number	22,194	65,290

Source: American Institute of Aeronautics and Astronautics, "AIAA National Membership Profile," New York, November 25, 1981.

Hawk, North Carolina. In comparison, a recent jet transport required over 6 million engineer-hours for design and development of the necessary aerodynamic surfaces, power plants, structures, and controls.

Aerodynamics   Modern aircraft develop incredible powers to carry tremendous loads faster than the speed of sound at altitudes where humans would perish quickly if unprotected. Every component of an airplane represents an engineering achievement in optimization. Aerodynamics engineers have responsibility for the efficient design of all surfaces that, in moving through the air, develop the desired forces of lift and the undesirable forces of drag. They devise and supervise the conduct of wind-tunnel tests, reduce and intrepret the results, and derive and test mathematical theories governing performance, stability, and control.

Aerodynamicists are interested in speeds ranging from practically zero in underwater craft and vertical takeoff aircraft to speeds of 5000 miles per hour for hypersonic planes and atmosphere-entry speeds of 50,000 miles per hour for vehicles returning to earth from space. Success in solving the problems in V/STOL (vertical/short takeoff and landing) aircraft has led to the development of short-haul transports capable of all-weather operation from short landing strips. The hypersonic aircraft may find application in high efficiency earth-to-earth or earth-to orbit transport.

Power plants   The first engine the Wright brothers built weighed about 13 pounds per horsepower; its modern piston-type counterpart weighs only one-tenth as much per horsepower. As speeds increase, propellers become less efficient; and as altitudes increase, jets become more efficient.

The modern turbojet is an air-breathing engine that works efficiently at high speeds. Air enters the inlet diffuser (see Figure 5-2) and is compressed at this location and in the compressor. Fuel burned in the combustor heats the air, which expands through the turbine. Some of the energy of the hot gas is used to drive the compressor, and the remainder, after further expansion in the jet nozzle, is converted into thrust. In some jets, an afterburner permits the addition of more heat and a corresponding increase in thrust. The development and refinement of the turbojet exemplify the work of the power plant engineer. Power is the major limiting factor for aircraft and space vehicles; power plant improvement is one of the aeronautical engineer's most challenging problems.

Structures   The structural engineer determines the loads to be carried by various components of the aircraft, conceives a framework that will carry the loads, and devises tests that will ensure the maximum safety and economy with the available materials. Striving for the ultimate in performance, working to extreme stresses with new materials in new shapes, and faced with challenging problems of flutter and vibration, the structural engineer must take advantage of every available theoretical and experimental tool to develop a light-weight, safe, and reliable vehicle.

Controls   Safe and efficient operation of modern aircraft requires devices that can sense, compute, and control more quickly and more

FIGURE 5-2
Schematic diagram of turbojet engine.

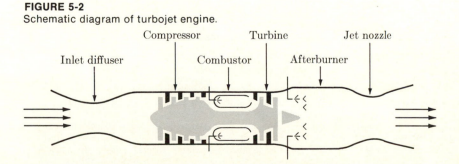

precisely than can a human being equipped by nature to operate at a walk at sea level. Every cubic inch under the skin of a jet fighter or space vehicle is jammed full of electrical, mechanical, hydraulic, and electronic gear to help the pilot navigate and maneuver. Indicator instruments provide some direct information, but more and more the control engineer must free the pilot from all but the judgment functions.

Air Transport

While most aeronautical engineers are employed by manufacturing companies or research and development laboratories, other engineers are needed by the airlines to develop specifications for new commercial aircraft, to evaluate new models, and to supervise operation, maintenance, and overhaul. In comparison to military operations, commercial airlines place more emphasis on economy, reliability, and comfort. Elaborate inspection procedures must be employed and efficient overhaul routines devised. Airport and terminal design, passenger handling, ticket reservations, food packaging, and in-flight entertainment become specialized engineering problems.

Astronautics

A spirit of adventure and a thirst for knowledge are the ingredients of our drive to conquer space. Already the *Sputnik,* the *Explorer,* and the *Columbia-Eagle* have taken their place in history alongside the *Niña,* the *Pinta,* and the *Santa Maria.* Engineers have always been adventurers into uncharted technical areas, but now they are turning their talents to exploring the clearly visible but largely unknown regions of our solar system and the vast universe beyond. Familiar technical factors take on new significance. Now the force of gravity is of concern in computing escape velocity; oxidation processes are considered in terms of thrust production; transmitters are expected to send back information from 350 million miles out in space, and a "living room" must protect its occupants from the effects of near vacuum, extremes of temperature, absence of oxygen, and presence of cosmic radiation.

Orbital mechanics The sun is a star around which nine planets including the earth revolve in near-circular orbits. The earth maintains its orbital

Simulation in the laboratory of conditions anticipated in spaceflight is an important tool in aerospace engineering. This simulator can impart to an astronaut five different types of motion simultaneously—roll, pitch, yaw, and vertical and horizontal motion. Here the NASA engineer takes a fix on the moon with a sextant in a simulator at the Kennedy Space Center Mid-Course Navigation Simulator. Aircraft simulators are also used to train commercial and military pilots. These simulations increase flight safety and decrease training costs. (*NASA*)

position by the precise balance between the gravitational force of attraction between itself and the sun and the centrifugal force due to its orbital velocity. Manufactured satellites move through space in accordance with the same laws that govern the motion of the moon about the earth and the earth about the sun. Generally, orbits are elliptical rather than circular and they are influenced by the primary body's deviation from a perfect sphere.

Propulsion   Every space vehicle consists of a propulsion system, a guidance and control system, a supporting structure, and a payload. The rocket engine used in most current space vehicles is a form of jet propulsion in which the fuel and oxidizer are self-contained; therefore, it can function outside the earth's atmosphere. In a typical rocket, kerosene and liquid oxygen are burned and the resulting hot gases are exhausted through a nozzle. As the gases are accelerated in the jet, the equal and opposite reaction (Newton's third law) provides a forward thrust. *Specific impulse* is an indicator of the total thrust capability of a propellant. Chemical propellants currently produce specific impulses of approximately 250 seconds; performance of approximately 400 seconds is theoretically possible. For long-term space missions to great distances, it is necessary to develop new propulsion systems with superior capabilities. Presently being studied are *nuclear fission,* whereby a reactor would be used to heat liquid hydrogen to obtain specific impulses of 1000 seconds; and *ion power,* whereby ions of propellant are accelerated to very high exhaust velocities by electric fields energized by a nuclear power plant or solar radiation, which promises even more efficient power for transporting long duration heavy payloads.[2]

Guidance and Control   In missile launching, the vehicle trajectory is frequently completely established during the propellant-burning period near the earth. The controls are programmed to start the missile off in approximately the right direction; the missile is then tracked by radar or onboard inertial measuring equipment, and the necessary corrections computed and communicated; vernier rockets bring it to the precisely correct direction and velocity; the power is cut off at the proper instant; and the missile coasts according to the laws of orbital mechanics.

In lunar landing flights of the *Surveyor* spacecraft, a midcourse correction was made on the basis of a computer trajectory analysis. Later, when the craft was about 40,000 feet over the selected landing area, a radar-controlled landing system took over for additional guidance and control. In the last two minutes, the craft was straightened up by vernier jets and slowed from 6000 miles per hour to a descent rate of 3 miles per hour by the main retro engine. At a height of 14 feet, the engine was turned off and the *Survey-*

[2]As Richard A. Roche has pointed out, "Mankind now occupies a spaceship of very advanced design which runs solely on solar power."

or dropped gently to the surface of the moon to begin taking its thousands of pictures.

Environment   Instrumented probes can obtain valuable data, but the complete knowledge we desire will require human judgment and ability to analyze observation. Humans must participate in person and, as Yuri Gagarin, John Glenn, and Neil Armstrong first demonstrated, the obstacles to safe launching and return can be removed by careful preparation. We have already learned much about the effects of "hi-G" forces during launching and reentry, weightlessness in space, extremes of temperature, oxygen concentration, X-ray and cosmic radiation, and bombardment by meteorites. We are now studying the ability of humans to perform complex functions over a long period of time in the close confines of a spaceship under conditions of stress. We are continuing to learn how to engineer life-support systems that will provide the environment in which men and women can live and function thousands of miles from earth for relatively long periods of time.

Life in Space

As a result of successful space exploration, life on earth will never be the same. *Space medicine,* evolved from aviation medicine, is concerned with the study, alleviation, and prevention of the hazardous consequences of manned flight outside the friendly environment of earth. Specialists study nutrition, respiration, circulation, sleeping, instruction receiving, and decision making under actual and simulated flight conditions. One of the great challenges is to establish a closed-cabin ecological balance so that a human's needs for air, food, and water can be met by recycling. On problems such as these, the engineer works with the physiologist, psychologist, and biochemist.

What would happen if an earth organism were adapted to Mars and then, generations later, brought back to earth? Could an earth plant be crossed with a plant from another planet? *Space biology* is concerned with the detection and study of extraterrestrial life on planets or in space. The discovery of extraterrestrial life would contribute greatly to our knowledge of the origin and evolution of life on earth. Engineers and biologists are now cooperating on the design of equipment for the biological analysis of extraterrestrial material. Astronautical engineers have opened new vistas to bioscientists leading to the rapid growth of the field of planetary science.

The Future of Aerospace Engineering

The rapid and continuous growth of aeronautics and the explosive development of astronautics give a clear indication of the extraordinary potential of

The GOES-E satellite provides improved meteorological data to the eastern half of the United States and Canada, all of Central and South America, and much of the Atlantic Ocean. Electrical engineers with diverse talents cooperate in the design of the antennae, transmitters, and computer relaying system utilized in the highly reliable data storage and retrieval system. Before encapsulation, technicians perform final servicing in a carefully planned series of operations designed to provide years of trouble-free service in a harsh environment. (*NASA*)

this fascinating branch of engineering. How far can it go? Look up into the heavens on a clear night; how far can you see? These are the boundaries of aerospace engineering.

ELECTRICAL ENGINEERING

The first commercial applications of electricity were in communication and illumination. Now the most useful properties of electrical energy are the ease and efficiency with which it can be generated and distributed and the precision and flexibility with which it can be applied and controlled. A unique property is the ability to store and process great quantities of information in the microscopic elements of a computer. These characteristics are being exploited by electrical engineers in an increasing number of diverse ways, as indicated in Table 5-5.

Communications

The ability to communicate with others far beyond the range of the spoken word represents a revolutionary step in human progress that has had cultur-

TABLE 5-5
DISTRIBUTION OF ELECTRICAL ENGINEERS
BY SUBBRANCH

Computer	20.5%
Power engineering	9.3
Communications	6.7
Industry applications	4.9
Circuits and systems	4.9
Electron devices	4.1
Control systems	3.8
Acoustics, speech and signal processing	3.7
Engineering management	3.6
Broadcast, cable, and consumer electronics	3.4
Aerospace and electronic systems	3.3
Engineering in medicine and biology	3.2
Microwave theory and techniques	2.7
Industrial electronics and control instrumentation	2.6
Total base number	230,062

Source: Institute of Electrical and Electronics Engineers, Inc. "Annual Report of the Secretary and Treasurer, 1980," New York, April 1981.

al, social, and political consequences. In sequence came the telegraph, telephone, wireless, and radio; talking motion pictures; the telephoto, and television; and satellite communication. Amazingly, a span of but 30 years covered the time from the first homebuilt crystal sets to the first color telecasts viewed by millions. Entire new industries have been established, the methods of warfare have been revolutionized, and the entire entertainment world has been reoriented to take advantage of new concepts and devices.

Communications engineering is firmly based on advanced physical concepts and mathematical techniques. Because radio waves and electrical signals are intangible, electrical engineers must work with abstract symbols based on indirect measurement and observation of natural phenomena. The results of their wizardry become apparent when the flip of a selector switch brings a color television program into your living room. Halfway around the world, light energy reflected from the performers strikes a camera tube and is converted into a multitude of electrical impulses. Each impulse is characterized by position, color, and intensity. This information, after amplification, is superimposed on a carrier wave and sent by cable to a transmitting station where it is converted to electromagnetic energy and radiated into space. Picked up by the antenna of a satellite, a small portion of the carrier signal is amplified millions of times and returned earthward. Huge antennas gather the weak signal, reprocess it, and send it to a relay station for reradiation.

At successive relay stations the radio signals are received, converted into electrical impulses, amplified, converted into electromagnetic energy, and reradiated. At the local transmitting station it is broadcast in all directions as electromagnetic energy. With the touch of your finger, this one signal is selected from the myriad of signals from radio stations, police calls, navigation aids, and amateur transmitters along with radiations from diathermy machines, electric storms, and passing meteors. Each little impulse is amplified, identified, and used to control a stream of electrons. These electrons strike fluorescent materials that give off light energy, thus forming a colorful reproduction of the original scene.

Power and Machinery

The capacity of electric-generating stations in the United States is about 800 million kilowatts and consumption exceeds 2 trillion kilowatt-hours annually. Generation, transmission, and distribution of this tremendous power are the responsibilities of electrical power engineers. They are concerned with locating hydroelectric, steam, diesel engine, and nuclear power plants, and specifying the engines, generators, and auxiliary equipment.

In large stations the power is generated at approximately 22,000 to 25,000 volts and is stepped up by means of transformers to as high as 1.2 million volts for efficient transmission over long distances. The efficient

Electrical engineers have successfully put the key component of an advanced superconducting generator through its most severe qualifying test, spinning its 13-foot-long, 1$\frac{1}{2}$-ton rotor at 3600 revolutions per minute while cooled to a frigid 452°F below zero. The tests confirmed that the experimental generator should be able to produce as much electricity as a conventional generator twice its size and weight. Following one of the tests, an engineer inspects the outer assembly of the rotor. This generator can produce 18 million watts of alternating current, enough electricity for a community of 20,000 people. (*General Electric*)

operation of individual power plants and the skillful shifting of load between elements to large transmission networks are supervised by electrical engineers. Near the load center, power is stepped down to lower voltages for safer distribution and easier control. The power is actually used at still lower voltages, usually 110 to 140 volts.

The end result of electrical energy is usually some other form of energy. Mechanical energy is supplied by efficient electric motors available in an almost endless variety of sizes, speeds, and operating characteristics. Thermal energy is obtained from easily controlled electric heaters and furnaces. Light energy in the infrared, visible, and ultraviolet portions of the spectrum is available from incandescent and fluorescent solids and gases. Magnetic energy is available for lifting tons of scrap iron or for operating sensitive galvanometers. Of the nearly 2.5 trillion kilowatt-hours of electrical energy produced each year, the largest single block is supplied to motors. Assuming an industrial rate of 4 cents per kilowatt-hour, mechanical power costs us about 32 cents per horsepower-day or about 4 cents per worker-day! It is no wonder that the electric motor has replaced the horse and human.[3]

Electronics

Electrons will move through a vacuum under the controlling action of electric or magnetic fields; and vacuum tubes, containing a source of electrons and two or more electrodes can be made to produce, detect, amplify, or rectify electrical signals. Electronics engineers are responsible for the design of efficient circuits employing electric elements that vary in size from the miniature receiving types sensitive to microwatts to the water-cooled transmitting tubes handling thousands of kilowatts. The *cyclotron* is a giant vacuum tube in which charged particles are accelerated to tremendous energies while moving in a circular path under the influence of a magnetic field.

Solid State Electronics   To move an electron out of a metal (into a vacuum, say) requires a large amount of energy ordinarily supplied by heating one electrode. Another characteristic of electronic tubes is that the vacuum is a poor conductor and energy is required to cause the flow of electrons. Within a metallic conductor, however, electrons move easily from atom to atom. A revolutionary development has been learning how to influence and control the motion of electrons within the molecular structure of certain carefully prepared semiconductors to perform many of the functions of vacuum tubes with a great saving in size and power requirement.

The transistor and the semiconductor diode were the first commercially important applications of the new technique. They made possible great

---

[3]"Weekly Output," *Electric World*, vol. 195, no. 7, July 1981, p. 15.

reductions in the size of existing equipment, but now integrated circuits (ICs) provide entirely new approaches to miniaturization, information storage, and energy conversion. An illustration is a device developed for measuring the radiation encountered by a satellite, converting the measurement to an electrical signal, and amplifying the latter for transmission to the earth. Using conventional components and miniaturization techniques, the device included 16 components in a volume of 4 cubic inches; an early IC performed all the designated functions yet occupied only 1/1000 cubic inch; now large-scale integration (LSI) permits construction of a system of a hundred such devices in that volume. There has been a corresponding decrease in cost from a few dollars for a transistorized circuit to a fraction of a cent for the LSI equivalent. The microprocessor, a versatile computer on a tiny silicon chip, has revolutionized electronic design. It is the basis for a new multibillion dollar industry in microprocessing and microcomputers, and VLSI (very large-scale integration) is already a $10 billion industry.

The study of semiconductors has made available materials with superior thermoelectric, optical, and magnetic properties that already have found practical application. For space power systems where weight must be minimized, engineers developed successful *solar cells* employing photovoltaic effects in silicon semiconductors. When *Surveyor I* dropped gently to the surface of the moon, thousands of solar cells mounted on outstretched paddles went into operation. Absorbing radiant energy from the sun to free electrons in the boron-diffused silicon, the power system generated 85 watts to operate its TV cameras and transmitters.

Quantum Electronics   In 1917, Einstein pointed out that an electron in a high energy state may be stimulated into giving up its energy to an electromagnetic wave of just the right frequency. In 1953, Townes (see Chap. 7) used this concept to achieve *microwave amplification by stimulated emission of radiation* in his *maser.* In 1958, Townes and Schawlow described a device that could extend the maser principle into the optical (or light) region; in 1960, Maiman succeeded in doing so with his pulsed ruby laser.

Since then quantum electronics has captured the attention of hundreds of organizations and thousands of engineers and applied physicists. Practical applications include piercing diamond dies, precision surveying, spotwelding retinas, and tunneling through rock. The laser beam is coherent and therefore can be focused into a fantastically small spot of extremely high intensity; this property can be used in tracking systems providing precise measurements of satellite position, velocity, and acceleration. The very high frequency of the beam can be modulated by other signals to provide a communications channel capable of handling simultaneously 100,000 telephone conversations or 100 TV broadcasts. As the beam is nearly

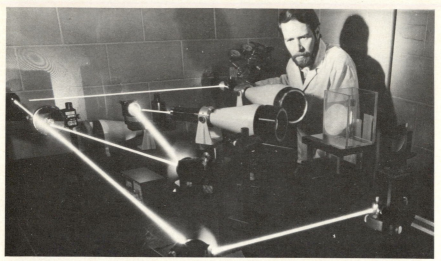

This apparatus provides special laser holograms that are used to learn more about heat transfer within containers used for transportation or storage of nuclear materials. The holograms show not only a test object—a model nuclear fuel cask, for example— but also how heat transfer is occurring within the object. Test objects are placed inside the transparent box (by the engineer's left shoulder) where a split laser beam is reunited. Sponsored by the Department of Energy, this project is being conducted in conjunction with work at Sandia to develop a cask for safely transporting spent nuclear fuel. (*Sandia Laboratories*)

monochromatic, it can be used to selectively alter atoms in a living cell or to enhance one chemical reaction while retarding another.

The availability of a coherent light source has created new interest in an old idea. If a coherent beam is used to illuminate an object (see Figure 5-3) and the scattered light from the object is combined with unscattered light from the same source, the resulting interference pattern or *hologram* contains information on the location of every point in the object. If coherent light is then projected through the two-dimensional hologram, an image of the object is recreated in space. The image is three-dimensional (3D); in observing a hologram, if you move your head you change your view of the object and you can *see around* an object in the foreground. Experimental 3D movies have been made using a succession of holograms. Since the interference pattern can be recorded photographically, it could be picked up by a television camera and transmitted. For displaying a 3D picture, you would need a continuously variable light-sensitive surface; this is the kind of problem engineers like to tackle.

Instrumentation

To *know,* you must be able to *measure,* and electronics has demonstrated the flexibility, the sensitivity, and the precision necessary to make the measurements so vital in research and operation. Because of the superior characteristics of electrical instruments, quantities such as temperature, pressure, speed, and flow rate are converted to electrical signals by transducers and then measured. In rocket research, for example, the important small changes in combustion-chamber pressures that occur in less than a millisecond must be accurately sensed by a mechanical element and converted to an electrical measurement.

Telemetering, measurement at a distance, is a familiar assignment for electrical engineers, but the transmission of observations from a satellite venturing millions of miles into interplanetary space introduced many new problems because of the distance, the environment, the great quantity of data to be transmitted, and the fantastic weight limitations. Nearly as difficult is the design of tiny electromechanical transducers to measure dimension and pressure changes within the human heart or Doppler systems to measure the rate of flow within an artery.

Controls

Another medical application of electronics is automatic control of the anesthesia level in a patient during surgery. The cortex brain wave, an accurate indicator of the depth of anesthesia, is sensed and integrated to obtain a

FIGURE 5-3
Typical setup for production of holograms.

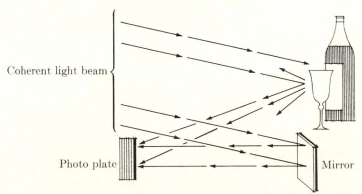

Coherent light beam

Photo plate   Mirror

measure of energy, and the energy level is continuously compared to the standard desired. An increase or decrease in energy operates a servomotor to open or close the gas valve to maintain the depth of anesthesia. This is an example of the important technique of *feedback control.*

The motor relieves human muscles, and simple devices can replace manual control of such elementary quantities as position or time. However, the automation of operations or processes requires comparing the actual output to the desired standard and taking the proper action to minimize any difference observed. In an autopilot, the actual spacecraft heading is sensed and fed back for comparison to the desired heading; the resulting error signal is amplified and the output is used to drive a hydraulic actuator to correct the setting of the vernier jets. The control systems engineer may select from mechanical, hydraulic, pneumatic, fluidic, or electric actuators; but because high amplification is required for stable operation, electronic amplifiers are usually used.

Computers

Carrying one step further the emancipation of humans from activities, the computer relieves us of one type of mental work—data processing. The slide rule and cash register were simple computers; data were fed in, computations were performed mechanically, and the results were read out. In contrast, a modern electronic digital computer can store 10 million numbers or instructions, perform a million multiplications involving 10-digit numbers, or print out 100 lines of results in 1 second. The electrical engineer is concerned with research in computer circuitry and materials, development of new concepts of storage and processing, design of faster and more reliable input and output devices, and application of computers to all types of scientific and industrial data processing problems.

The computers used in business for banking, accounting, payroll, and inventory perform relatively simple operations on great quantities of data. The computers that serve engineers and scientists in problem solving and research perform more complex operations on smaller inputs. Now, a major engineering application of computers is in controlling machines and processes, or even entire plants. For example, the complete control of a refinery is a suitable task for a large digital computer in conjunction with several specialized computers directly connected to processing units.

The ultimate goal is an electronic data processing system with the complexity and compactness of the human brain, plus the speed of electronics. Such a device would be able to recognize a photograph, something any child can do, but beyond the capability of any present computer. The human brain contains about a trillion neural elements and an equivalent computer would have to employ units a thousand times smaller than the

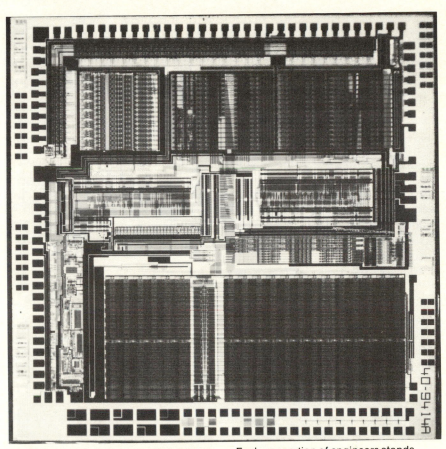

Each generation of engineers stands on the shoulders of those who have gone before. Combine our understanding of semiconductor physics, our skill in computer-aided design, our knowledge of computer architecture, and our precision in process control with the desire to solve a challenging problem and the result is the "superchip" shown. On this $1/4$-in square bit of silicon, little larger than older single-transistor capsules, one-half million transistors function as a microprocessor with input and output ports, instruction registers, timing and control networks, and bus controllers. With a custom program, memory, keyboard, and display, this microcomputer will enable engineers to solve problems leading to a new round of product development.
(*Hewlett-Packard*)

microminiature elements now in use. Electrical engineers will participate in this development through progress in microelectronics, switching theory, pattern recognition, and information storage and retrieval. Computers of the future will not only recognize and process information, but will also adapt to experience (in other words, "learn") and exercise judgment based on previous experience.

The Future of Electrical Engineering

It would take many years for engineers to exploit fully the potentialities of electrical concepts and devices available now. But researchers are not waiting for engineers to catch up! A conservative prediction is that within the next 10 years there will be more new developments than in the previous 100 years in electrical engineering. Here is an area with great opportunity for those with aptitude, creativity, and ambition.

CHEMICAL ENGINEERING

Chemistry has to do with the composition and changes in composition of substances and with the preparation, separation, and analysis of substances. Chemical engineering is concerned with the industrial-scale manufacture of substances from raw materials through controlled chemical and physical changes. Chemical engineering is closely allied with, and overlaps considerably, ceramic, petroleum, metallurgical, energy conversion, and sanitation engineering. The emphasis in this discussion will be on the work of those who call themselves chemical engineers, the industries where chemical engineers find employment, and the processes whereby substances are manufactured efficiently and economically.

The interests of chemical engineers are reflected in the technical specialties and employment of members of the American Institute of Chemical Engineers (AIChE) (Table 5-6).

Chemical Industries

Chemical engineering includes developing new products, designing new and improved processes, and operating plants efficiently; therefore, the chemical engineer may work in a laboratory, an office, a pilot plant, or a full-scale plant. Chemical engineers are useful in a great variety of enterprises; about two out of three are employed in industries manufacturing such products as chemicals, petroleum, plastics, and foods.

Chemicals   The *heavy* chemicals include various acids, alkalies, salts, and other substances used in great quantities in manufacturing and process-

TABLE 5-6
DISTRIBUTION OF CHEMICAL
ENGINEERS BY JOB ACTIVITY

Chemical engineering	4.5%
Data processing	2.6
Elastomers	3.4
Engineering research	3.3
Flow of heat	4.1
Fluid mechanics	3.9
Instrumentation and control	2.0
Management	11.8
Mass transfer	3.7
Plant design	4.2
Pollution control—air	2.1
Pollution control—solid wastes	3.0
Process design	9.1
Process development	9.5
Research	3.0
Other	3.3
Total base number	49,077

Source: American Institute of Chemical Engineers, "Quarterly Statistical Report," New York, November 1981.

ing. The *fine* chemicals include pharmaceuticals, cosmetics, insecticides, and photographic materials. Presently, the plastics are of great importance because they are cheap, light, easily molded, and chemically resistant. Rayon and the newer synthetic fibers available since 1940 have revolutionized the clothing industry with materials softer than silk, warmer than wool, and cheaper than cotton.

Petroleum   Some of these new fibers are produced from by-products of the petroleum industry. In fact, petroleum is now recognized as a versatile raw material for manufacturing a great variety of important products—the petrochemicals. The fuels, lubricants, and solvents made from petroleum have established themselves as essential parts of our industrial operations. Other valuable derivatives include detergents, soaps, and the polyethylene and polystryrene plastics.

Rubber   The rapid adoption of the pneumatic-tired automobile made the United States the world's largest consumer of rubber products. Rubber is made from the gum of a tropical tree, and its natural supply is limited. World War II stimulated the demand for a substitute; the development of synthetic rubber marked a milestone in chemical history because it involved changing molecular structure to create an entirely new substance.

Foods Many of the activities of the engineer are aimed indirectly at improving the world's food supply in quantity, quality, and economy. The chemical engineer works directly in producing and processing food products. As a result of his work, previously discarded cotton seeds are now turned into fertilizer, animal feed, and valuable vegetable oil. Corn now yields a long list of valuable materials including dextrose, glucose, corn syrup, and corn oil. Hydrogenation of vegetable oils transforms them into fats, which may be used in place of the more expensive butterfat and animal fat. Preservation of foods by cooking, spray drying, fast freezing, and irradiation are important contributions.

Biochemicals The laboratory development of antibiotics like penicillin, hormones like cortisone, and immunizers like poliomyelitis vaccine held tremendous promise for society if these remarkable substances could be produced in great quantity at reasonable cost with all the necessary purity and precision. The design and operation of full-scale plants to reproduce the biochemical and bacteriological processes of the research laboratory were outstanding chemical engineering achievements. In attempting to improve world health in the face of a population explosion, our ability to make universally available new drugs, contraceptives, fertilizers, and insecticides will be a critical factor. The chemical engineer's greatest contribution is in adapting test-tube reactions to mass production so that the research chemist's work benefits all.

Unit Operations and Processes

The manufacturing processes employed are made up of combinations of a relatively few basic chemical and physical processes called *unit operations.* These can be grouped into three general classifications.

Separation Processes Materials handling, pumping, mixing, shredding, crushing, grinding, sizing, filtering, and settling all involve *mechanical action.* The machines that perform these operations must be cleverly contrived to achieve the desired result efficiently, and they are usually power driven for quantity output. In addition to designing the equipment, the chemical engineer must provide accurate instruments and automatic controls.

Over and over again the chemical engineer faces the problem of separating the desired product from that which is undesired. *Physical separation* may be achieved by evaporation, absorption, adsorption, distillation, frac-

A construction supervisor checks the progress of a multimillion dollar refinery investment in Louisiana. This maze of piping, columns, towers, and tanks will efficiently convert raw materials into valuable ammonia. The various vessels provide optimum conditions for physical separation, chemical reactions, heat conduction, mass transfer, and fluid flow on a continuous, closely controlled basis. Processes like this, developed in the laboratory using glass apparatus, are verified in pilot plants before full-scale steel and concrete plants are constructed. (*Monsanto Company*)

tionation, crystallization, and precipitation. The first operation of these processes in the research laboratory may be carried out in small glass units, whereas the final production plant may tower a hundred feet in the air and represent a million-dollar investment.

Chemical Reaction   In many processes, a chemical change must be produced as in smelting ore or making soap. Other "unit processes" involve oxidation, reduction, hydrogenation, chlorination, and polymerization. These reactions are carried on in tanks, furnaces, retorts, towers, and other units especially engineered to produce the desired result at a minimum cost. In creating new detergents that would not cause trouble in sewage systems, chemical engineers used radiation to stimulate chemical reactions selectively. They are now studying biosynthesis processes that could use microorganisms to convert low-cost petroleum into valuable protein food products.

One of the most challenging problems is how to make a reaction proceed rapidly; solving such a problem requires a knowledge of chemical kinetics and catalytic action. The rate at which reactants are used up or products are formed is particularly important in organic chemistry or biochemistry. In general, a given set of reactants may yield many different products. The relative rates of the competing reactions determine the product that is actually obtained. In ordinary thermal reactions, the energy necessary for chemical change is provided by collisions with high-velocity molecules. In other cases, a chain reaction provides its own energy input.

Heat and Mass Transfer   Frequently, the output of a given plant may be increased by carrying on the reaction at high temperature. Some desired reactions will take place only within narrow temperature limits, and cooling may be necessary. *Heat transfer* involves conduction, convection, and radiation. The chemical engineer works with heat exchangers, furnaces, evaporators, condensers, cooling towers, and refrigeration units.

A common assignment for a chemical engineer is to create a system for making, by the ton, a material that some chemist made by the ounce in a laboratory. To merely build a kettle that will hold a ton of chemicals is not the solution. Continuous processes, usually involving *fluid flow*, are generally more satisfactory. The chemical engineer must take advantage of knowledge of laminar and turbulent flow and how required flow rates are reflected in pump and compressor specifications.

In addition to flow processes, *mass transfer* occurs by means of free convection, molecular diffusion, interphase transfer, absorption, humidification, and drying. Frequently, the effectiveness of an operating unit depends on the rate of *momentum transfer* between moving fluids—for example, as displayed by the violent mixing that occurs in combustion.

Control and Optimization

In setting up a manufacturing plant, the chemical engineer selects the individual processes and arranges them in the proper sequence. In any commercial installation, piping presents many problems; and furthermore, there are always structural problems. Of great importance are the process controls and instrumentation that enable the engineer to accomplish the desired result and keep track of the performance of each unit.

In addition, the overall system must perform in an optimum way to yield the best product mix from a given feedstock at the lowest cost. Systems analysis and optimization techniques are common tools of the modern chemical engineer, and analog and digital computers are widely used to ensure that technical and economic objectives are met.

The Future of Chemical Engineering

As one of the youngest of the major branches of engineering, chemical engineering possesses a tremendous growth potential. Probably there has been more industrial research in this branch than in any other field. It is not uncommon for a major company to point out that one-half its current sales involve products that were either unknown or not commercially available a decade ago. Challenging opportunities are offered by the advent of space travel requiring high-energy fuels and life-support systems for hostile environments, the rapid advances in nuclear engineering and the associated radiation effects, the population explosion and the need for increased food supply and pollution control, and the increasing importance of a practical method for seawater conversion. The American chemical industry was established as a result of World War I and greatly stimulated by World War II. Its continued growth is certain under the stimulus of the never-ending war for better health and a higher standard of living.

AGRICULTURAL ENGINEERING

The population explosion has focused attention on our basic need to provide ever-increasing amounts of food and fiber. Engineering takes advantage of energy, methods, and mechanisms to multiply the effectiveness of humans, and agricultural engineers apply fundamental knowledge and modern practice to improving agricultural productivity. Their knowledge embraces both biological science and physical science, and their practices are drawn from all other branches of engineering. The 1981 American Society of Agricultural Engineers' profile of its 8800 members indicated that about one-third were employed by manufacturers of agricultural equipment (27 percent) and components (5 percent). About 4 percent were employed in food and feed processing, 16 percent were in government, 21 percent were in education, and 11 percent were in other areas; ASAE members indicated that their primary technical interests were distributed as follows: electric power and processing (51 percent), soils and water (27 percent), structures and environment (11 percent), power and machinery (9 percent), and food engineering (3 percent).

The agricultural engineer specializing in *electrification* applies electrical energy in producing, processing, and inspecting products. New photoelectric sorters, for example, can look inside eggs and fruits, and determine eating quality. The specialist in *farm structures* designs barns, shelters, silos, driers, and processing centers. To provide the ideal environment for poultry or cattle, an understanding of animal husbandry must be combined with knowledge of heating, ventilating, and control.

To meet the increasing demand for food, unproductive lands must be reclaimed. *Soil and water conservation* engineers bring water to arid land and improve the land's effectiveness by grading, terracing, and constructing ponds. In very porous soil, agricultural engineers have placed underground asphalt layers to hold moisture in the root zone. The mechanization of machinery and the development of new equipment have greatly benefited the farmer. Specialists in *power and machinery* recently developed a mechanical tomato harvester that increases the picking capability of its crew tenfold; plant breeders cooperated in the project by developing a tomato plant that stands upright and ripens all of its fruit at the same time. Solar energy is harnessed to dry grain, corn, and other products.

Engineering efforts in the various specialties of agricultural engineering have been traditionally production-oriented, that is, application of engineering principles to improving the productivity of the farmer. Another new and expanding area of agricultural engineering is *food engineering*. This field involves the agricultural production of food after it leaves the farm gate, or the preparation of food for human consumption. This includes the design and development of equipment along with methods and processes for canning, roasting, drying, freezing, and packaging the food products.

In contrast to most countries of the world, the United States produces more food than it needs. In the past 20 years, U.S. farm production per worker-hour has increased by 150 percent; one U.S. farm worker now produces food for seventy-eight people—fifty-two U.S. residents and twenty-six foreign residents.[4] This spectacular increase is due to technological improvements that are the envy and the hope of other nations. The agricultural engineer has an unusual opportunity and responsibility to serve humanity.

NUCLEAR ENGINEERING

The availability of almost limitless energy through atomic fission and fusion stimulated imaginative persons in all branches of engineering. Electrical engineers saw the possibility of generating tremendous amounts of inexpensive electrical power. Civil engineers conceived of constructive applications of the explosive energy of the bomb. Aeronautical and mechanical engineers thought of possible nuclear propulsion systems. Chemical and agricultural engineers and medical people envisioned applications of the accompanying radiation. As knowledge of nuclear processes expanded and as application techniques developed, however, nuclear engineering became a separate branch.

Although the membership of the American Nuclear Society includes many people from nuclear science and allied fields, nuclear engineers constitute the majority of the membership. In 1981, the American Nuclear So-

---

[4]1980 USDA statistics.

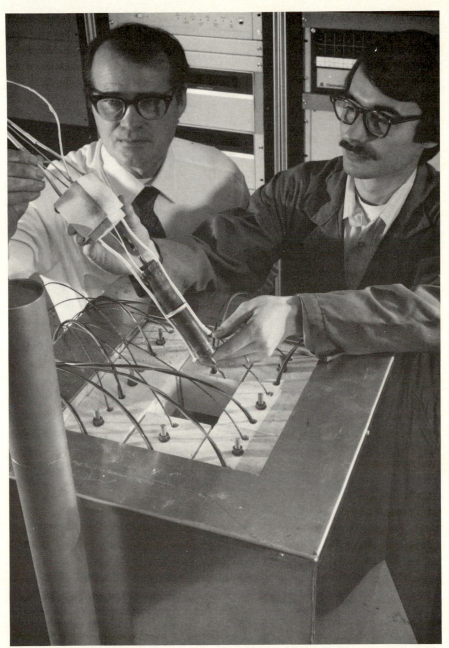

A superbattery for utility energy storage and electric vehicle propulsion applications is being developed jointly by American and British companies. Making use of sodium and sulfur reactants and a "beta-alumina" ceramic electrolyte, the new beta battery promises to outperform and outlast conventional lead-acid batteries by a wide margin. Preparing a laboratory-size sodium-sulfur cell for tests are an electrochemical engineer and a test technician. (*General Electric*)

ciety indicated that about half of its members were in management; 15 percent were in high level positions (e.g., president, owner), 32 percent were in middle management (e.g., supervisor, manager), 29 percent were in engineering positions (e.g., nuclear engineer), 12 percent were in science (e.g., physicist, chemist), 5 percent were in education (e.g., professor), and 5 percent were in other positions. They were working for a wide diversity of employers including the federal government (24 percent), construction-consulting firms (24 percent), utilities (18 percent), manufacturing (17 percent), education institution (6 percent), and private research laboratories (6 percent).

Nuclear Power

The first nuclear central station started operation in 1957, and within a decade nuclear power became economically competitive with fossil fuel power in most of the United States. France plans to produce half of its electricity using nuclear power by 1987. Almost one-half the new electric-generating capacity now being ordered in the United States is nuclear, and it is estimated that by the year 2000 one-third of all electrical power will be generated in nuclear power plants. The use of nuclear power will extend our energy resources, reduce air pollution, and reduce the cost of electrical energy.

Most nuclear engineers today are involved with the design and operation of safe nuclear power plants. These people work for the utilities that operate the power plant, the companies that design and build the reactors and power plants, the companies that supply special services for the operation of power plants, and the Nuclear Regulatory Commission that inspects and licenses the whole industry.

In a common type of commercial nuclear reactor, heat is generated by the fission of an isotope of uranium, uranium 235. The heat is removed by boiling water and the resulting steam is used to drive an ordinary turbine. Only 0.7 percent of natural uranium is uranium 235; breeder reactors are being developed to permit utilization of the remaining uranium 238.

In the fast breeder reactor shown in Figure 5-4, some of the excess neutrons (beyond those required to continue the chain reaction) are used to convert uranium 238 to fissionable plutonium, plutonium 239. The net result is that more fissionable fuel is created than is used and the cost of fuel becomes extremely cheap. Many nuclear engineers are struggling with the difficult problems in developing competitive fast breeder reactors. The fast neutrons do not interact with the fuel as readily as slower neutrons, so fuel cores must be compact; this creates problems in structure, handling, and control. A cheap coolant with good thermal properties that does not slow down neutrons is needed. Sodium meets these requirements, but special precautions must be taken as it reacts violently with water or air.

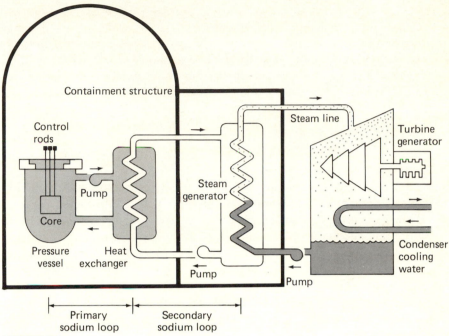

FIGURE 5-4
Schematic of nuclear power plant. Shown is a liquid metal fast
breeder reactor (LMFBR).

Fusion

A process which is the opposite of fission is *fusion,* in which two light nuclei
are fused together with a consequent release of energy. Such thermonuclear
reactions utilize plentiful hydrogen and would represent an almost infinite
source of energy if they can be made to work safely and economically.
Nuclear engineers and physicists are working on many different ways of fus-
ing hydrogen; and nuclear engineers are studying ways to convert the
energy released into useful forms of energy such as heat and electricity.
Japan is pouring $1.2 billion into nuclear energy research and hopes to
have a fusion reactor operating by 1990.

Space Program

Nuclear engineers will play an important role in space exploration because
nuclear energy sources may be compact and long-lived. Deep-space probes
which operate away from the sun cannot use solar cells to supply their
energy. Therefore, probes such as the Pioneer and Voyager series used
nuclear heat sources to keep the components from freezing and to produce

the electricity to operate the instruments. For deep-space voyages, the most suitable engine may be a nuclear-electric rocket in which electrical energy is generated and used to accelerate charged particles. The low thrust of perhaps only 20 pounds, provided by the reaction force, could over a long time build up a velocity much greater than that possible with a chemical rocket and thereby greatly shorten the travel time to distant planets.

Radioisotopes and Radiation

The use of radioisotopes as a diagnostic tool is rapidly expanding. Radioisotopes are used to check the performance of body organs such as the thyroid and the flow of blood in the brain, and are a useful tool in diagnosing certain problems in the heart. Scientists routinely set up radioisotopes to follow the paths of chemicals in organic tissue during medical research and during animal and plant development. Radiation is used to sterilize medical supplies and is an important tool in fighting cancer. In the future, radiation will also be used to preserve food without refrigeration. Nuclear engineers are concerned with all phases of producing the radioisotope, the development of the detection instrumentation, and the design of radiation sources and shielding.

The Future of Nuclear Engineering

Nuclear technology is still relatively new and many important developments lie ahead. The accident at Three Mile Island increased public concern regarding nuclear power plant safety as well as environmental pollution, although nuclear power is already regarded by most energy experts as the safest and least polluting source of energy available. The nuclear engineering profession is concerned with learning from the Three Mile Island incident and making the existing nuclear power plants and those of the future even safer. The expanded use of radioisotopes and radiation in research and medicine will lead to earlier discovery of medical problems and new and better ways of fighting disease. The use of radiation to preserve food may decrease the cost of foods through the use of fewer pesticides and less refrigeration. While breeder reactors will give us an abundant supply of energy, the long-term goal is a fusion reactor that will make available the limitless energy in ordinary seawater. By making available vast amounts of economical electric power, nuclear engineers will contribute to a worldwide economic and social revolution.

ARCHITECTURAL ENGINEERING

The architect encloses space to provide the optimum environment for humans to live, work, or play; he or she achieves beauty through use of dimensional form and spatial relationship. Architectural engineers contribute their knowledge of building design, materials, structural elements, mechanical and electrical equipment, acoustics, and illumination to provide buildings that are safe and efficient as well as aesthetically satisfying.

The role of the architectural engineer becomes more important as the building increases in size and complexity. In high buildings, the strength of the structure becomes of primary importance, and structural analysis for

The skyscraper, an American innovation, took advantage of new engineering knowledge of materials and fabrication techniques to solve an old problem of providing space in urban areas. Underground buildings can also be space and energy efficient if properly designed and constructed. This attractive, new headquarters addition for a major insurance company has these features. (*Leo A. Daly, Architectural Engineer*)

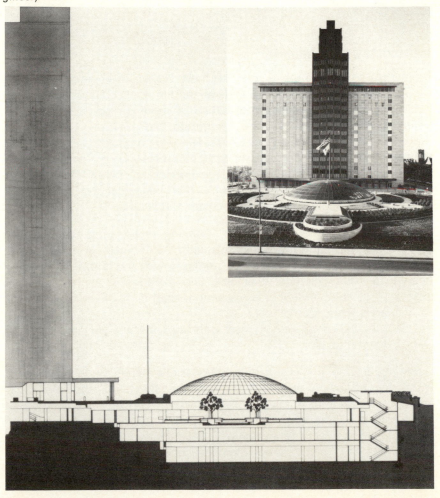

wind loads, and earthquake stresses may be performed with the aid of a computer. In office buildings the structural, air conditioning, plumbing, lighting, communication, and transportation systems must be carefully integrated in design and coordinated during construction. Design decisions and construction supervision require the technical knowledge of the architectural engineer.

Architectural engineers may work as private consultants, in partnership with an architect, for a contracting firm, or for a government agency. The present emphasis on urban problems focuses attention on the quality of housing as a key factor in determining whether city dwellers are content. Architectural engineers have an opportunity to contribute to human welfare by providing satisfactory housing at low cost through the development of more efficient structures, more durable materials, or more economical construction methods.

BIOMEDICAL ENGINEERING

Biomedical engineering is a relatively new branch of engineering which supports the rapidly developing medical and life science areas. A biomedical engineer uses engineering expertise to solve biological and health related problems. These include the design and production of artificial limbs and organs as well as instrumentation and computers to monitor patients and to optimize hospital systems. To prepare for this occupation, a background in electrical, mechanical, industrial, or chemical engineering is usually needed along with education in the biomedical and life sciences. There are three subbranches of biomedical engineering:

Bioengineering—concerned with applying engineering to the nonmedical biological and life sciences (usually the nonmedical systems)

Medical engineering—concerned primarily with the utilization of engineering in instrumentation, diagnostic and therapeutic devices, artificial organs, prosthetics, and other equipment needed in medicine and biologically related fields

Clinical engineering—concerned with applying engineering to health care delivery systems

Bioengineering involves many traditional disciplines, on the one hand, in engineering and physical science and, on the other hand, in the biological and life sciences. Bioengineers apply engineering concepts and technology to advance our understanding of nonbiological systems; NASA refers to this body of knowledge as biotechnology. Bioengineers are concerned with maintaining and improving environmental quality and protecting human, animal, and plant life from toxicants. Bioengineers apply engineering principles to biological systems including the production and processing of foods

The electronics design engineer holds a nonimplantable insulin-pump prototype designed to deliver predetermined amounts of insulin into a diabetic's system. The goal of this engineering laboratory–medical-school project is a smaller system that can be implanted into the abdominal cavity of a diabetic. (*Sandia Laboratories*)

and fibers, the study of animal migration, and a wide range of biological, environmental, and ecological studies.

Medical engineering uses engineering concepts and technology related to the structure, function, and pathology of the human body. Medical engineers develop instrumentation, materials, diagnostic and therapeutic devices, computer systems, artificial organs, and other equipment needed in biology and medicine. Medical engineering specializations include prosthetic devices such as limbs, bones, joints, and cardiovascular materials as well as artificial organs and assist devices such as pacemakers and defibrillators.

Clinical engineering uses engineering concepts and technology to improve health care delivery systems in hospitals and clinics. The diverse challenges of clinical engineering reward the men and women who have interests in electronic and mechanical devices and in the life sciences. Clinical engineers are members of the health care team and work closely with physicians in surgical and intensive care units and with rehabilitation therapists in designing prosthetics and prosthodontics. There is a wide variety of career specializations available within clinical engineering including health care delivery systems, hospital safety, and monitoring of postoperative, intensive, coronary, and rehabilitation care systems.

Although biomedical engineering is expected to grow faster than most occupations, the number of openings in this small but demanding engineering field is not expected to be large. Bioengineers in the future will require multidisciplinary and interdisciplinary education and training beyond the B.S. degree.

COMPUTER ENGINEERING

Engineering, a dynamic profession, continues to branch out as new specialties achieve the status of separate disciplines. Rapid advances in the processing of information by machines have focused attention on a broad group of principles and techniques generally referred to as *computer science.* Actually it includes very little science as we have defined it. Instead, it is a partnership of mathematics and engineering. The logical processes employed and the basic structure of a computer are purely mathematical; the successful computer engineer must think like a mathematician. The operation and realization are purely engineering; advances in computer technology require the solution of extremely difficult engineering problems.

In the *analog computer,* information appears in the form of numerical values represented by continuously varying quantities such as displacement or voltage. Electronic circuits can act as multipliers and integrators, and analog computers are particularly useful in rapidly solving sets of differential equations representing physical systems. Because it is so fast and so responsive to the operator's suggestions, the analog computer is an important tool for the development or design engineer. By changing inputs, initial conditions, and parameter values, the engineer can rapidly determine the effects of such changes and quickly arrive at an optimum design.

In the *digital computer,* information is in the form of discrete numbers or digits. In the binary system there are but two digits, 0 and 1 corresponding to the OFF and ON positions of a switch. The electronic digital computer is essentially a high-speed machine for counting digits in a logical way in accordance with instructions. As indicated in Figure 5-5, instructions and data are accepted by an input device. The control unit interprets, selects, and executes the instructions. Instructions, data, and intermediate results are stored in the memory. A processing unit includes logic circuits

FIGURE 5-5
Functional components of a digital computer.

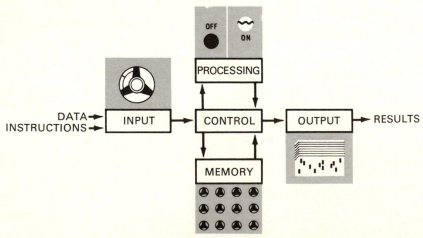

for adding, subtracting, multiplying, dividing, and other arithmetic operations. The results are presented in printed, graphical, or punched form by the output device.

Historically, the last half of the twentieth century may well be called the computer age. Certainly the high-speed general purpose digital computer is revolutionizing society to the same extent as the steam engine once did. The banking, publishing, merchandising, and manufacturing industries; the medical, legal, teaching, and engineering professions; and most branches of government are profoundly affected by the computer. Not only does the computer perform the calculations necessary for solving problems of increasing complexity; in addition, it provides new insights into old problems, it simulates the operation of existing and proposed systems, and it assists humans in the learning process. Also, by developing its own artificial intelligence, it forces us to redefine human intelligence and the role of humans in thinking and decision making.

Numerical analysis, machine languages, algorithms, and automata theory are highly mathematical in nature. The computer engineer will be primarily concerned with advances in programming, computer architecture, logical design, memory units, and output devices. Along with a strong mathematical background, particularly in discrete mathematics, the engineer planning on a career in this field will want to study switching theory, digital circuits, electronic devices, coding, and information theory. If computer capability continues to increase tenfold every three years, within the lifetime of today's engineering students computers are going to approach some of humankind's intellectual abilities—and surpass other abilities. The well-prepared computer engineer will have an opportunity to participate in some of the exciting discoveries and important developments.

SUMMARY

The engineering field is broad and diverse. It is concerned with modifying molecules at one extreme and working with tremendous masses of concrete at the other; it embraces destruction and creation; it includes use of minute quantities of electromagnetic energy plucked from the atmosphere as well as attempts at harnessing the vast energy of the sun. There is a branch of engineering to match almost any interest. However, all the specialized engineering subdivisions are unified in method and objective—"applying science to the optimum conversion of natural resources to the benefit of man."

REFERENCES

1 American Society of Mechanical Engineers, "Definitions of Occupational Specialties in Engineering," New York, 1951.

2 "Weekly Output," *Electric World,* vol. 195, no. 7, July 1981, p. 15.

3 Potvin, A. R., Long, F. M., Webster, J. G., and Jendrucko, R. J., "Biomedical Engineering Education: Enrollment, Courses, Degrees, and Employment," *IEEE TRANS. Biomed. Eng.,* vol. 28, no. 1, 1981, pp. 22–27.

4 The American Society of Agricultural Engineers, "1980–81 Agricultural Engineers Yearbook," 27th ed., St. Joseph, Michigan, 1980.

5 American Society of Civil Engineers, "Directory 1978 American Society of Civil Engineers," New York, October 1978.

6 American Society of Mechanical Engineers, "ASME Membership List," New York, July 1980.

7 American Society of Mechanical Engineers, "ABC Report, Geographical Area," April 30, 1981.

8 American Society of Mechanical Engineers, "Inter Office Memo" (Technical division counts as of 9/81 update), September 30, 1981.

9 American Institute of Chemical Engineers, "Quarterly Statistical Report," New York, November 1981.

10 American Society of Civil Engineers, "Official Register," New York, 1981, p. 99.

ASSIGNMENTS

1 List three branches of engineering that would participate in the design of a communication satellite and describe a specific engineering problem that might be assigned to an engineer from each branch.

2 Repeat Assignment 1 for the planning of a water pollution control program.

3 Repeat Assignment 1 for a research project aimed at improving food production in Mexico.

4 Select a specific engineering device, structure, or system that probably required the cooperation of civil and chemical engineers, and describe the contribution of each.

5 Repeat Assignment 4 for a project requiring the cooperation of materials and electrical engineers.

6 Repeat Assignment 4 for a project requiring cooperation of mechanical and industrial engineers.

7 Repeat Assignment 4 for a project requiring cooperation of nuclear and computer engineers.

8 Select a branch of engineering that interests you, and interview an engineer engaged in that work. Discuss with him or her the future of that branch of engineering. In a brief report, indicate the engineer's name and position and list five new developments to be expected or problems to be solved in that branch in the near future.

9 From the current periodicals in your school or public library, obtain information about a recent engineering project or achievement and answer the following questions:

a What is the general nature of the project?

b In what ways does this represent an advance or an innovation?

c What branches of engineering are represented by the engineers engaged in this work?

10 There are some engineering subbranches that may soon gain full branch status. These include (a) ocean engineering, (b) environmental engineering, (c) biomedical engineering, (d) solar engineering, and (e) system engineering. Select one of these and describe (in a one-page report) its origin, present activities, and possible future.

11 In cost and human resources, landing a human on the moon may have been the largest engineering project in history, far surpassing the construction of the pyramids or the development of the atomic bomb. One justification for this huge investment was that the new technology would be transferred beneficially to life on earth. Describe three specific benefits the space program has had for: (a) civil engineering, (b) materials engineering, (c) mechanical engineering, (d) electrical engineering, (e) chemical engineering, and (f) agricultural engineering.

6

FUNCTIONS OF ENGINEERING

The branches of engineering indicate what the engineer works *with*. For example, when you hear the words "civil engineer," you probably think of bridges, buildings, and highways. "Electronic engineer" immediately produces a mental image of transistors and integrated circuits. But in your mental picture, what is the engineer *doing?* What would you be doing if you had a career in one of these branches of engineering? The devices, products, and structures that have been described are the *results* of engineering. Engineering itself is essentially a mental activity; you very seldom *see* engineering. To complete your picture of engineering you need to know what engineers *do*. This can be described best in terms of functions.

Another reason for emphasizing the functional classification is that while branches usually relate to the engineer's interests, the functions tie in more closely with his or her aptitudes and training. A certain type of individual is successful in research, whether it is in the field of chemistry or electronics. The training required to design a bridge is basically the same as that necessary to design an aircraft fuselage or a microwave antenna. A sales engineer must be prepared to make recommendations from a company's entire line of products, whether they be electrical, mechanical, or chemical in nature.

Listed in order of decreasing emphasis on scientific concepts, the major engineering functions are *research, development, design, production, construction, operations, sales,* and *management.* This is a finer classification than usually occurs in industry. Research and development are frequently combined in a single department. In some organizations, the same persons perform design and construction. Sometimes production engineering responsibilities are broadened to include operation as well. The more precise definitions are used here to emphasize to you, the student, the small but significant differences in responsibilities and requirements. Other important functions of interest to prospective engineers are consulting and teaching.

To provide an understanding of what engineers do, each major function is defined briefly and then discussed in detail with illustrative examples. To help you in your consideration of possible careers, the personal qualifications for work in each function are listed and the most desirable type of training is indicated. Emphasis is placed on the concept of engineering as a broad spectrum of activities offering a variety of careers rather than as a single, narrowly defined activity.

RESEARCH

Research has been described as a "blind man in a dark room looking for a black cat that probably isn't there." This statement highlights two important aspects of research work. First, since research scientists or engineers are working on the frontier of knowledge, they are feeling their way over unfamiliar ground, usually with inadequate tools and techniques. Second, in their groping they must expect many failures before they attain success. In some cases success may be of a negative kind; many months of work may be required to prove that a certain theory is *not* true or that a certain method will *not* work.

Relationship of Engineer to Scientist

There is a close relationship between the research engineer and the research scientist. However, their objectives are different; the aim of the scientist is to know—to discover truth for its own sake—whereas the engineer, by our definition, is guided in his or her search by the desire to satisfy a human need. For example, a physicist may be interested in discovering the principle of atomic behavior underlying the phenomenon of superconductivity; the engineer working in the same field concentrates his attention on what happens in materials that could be used as cryogenic memory elements in a computer. A chemist may make her contribution by carefully measuring the properties of a new fluorocarbon plastic; the research chemical engineer focuses his effort on understanding and enhancing the heat-resisting and nonsticking characteristics with a view toward exploiting them, as in Teflon. Sometimes the distinction is expressed by calling one *basic* research and the other *applied* research.

Examples of Research Problems

As you read this, thousands of American scientists and engineers are engaged in the mental and physical activity called research. To illustrate the work they are doing, let us consider some typical research problems.

Aerodynamics   What determines the distribution of pressure and velocity over an airfoil section at supersonic speed? How are these factors affected by temperature, density, and viscosity of the air; also by roughness, temperature, and angle of the airfoil? What determines the magnitude of the shock wave and the resulting sonic boom? Most research work is directed toward answering questions of this type through the use of advanced tools and techniques. In the field of aerodynamics, tools include wind tunnels that will accommodate model or full-scale aircraft, and sophisticated apparatus for observing flow patterns and shock waves. Another powerful tool is mathematics, which makes it possible to express fundamental physical relations and to reduce masses of data to precise conclusions. By using mathematical principles, it is possible to determine from tests on small models just how full-scale planes will perform. Computers are used to simulate the dynamic processes occurring in airflow at supersonic speed.

One standard technique in engineering work is the experimental method wherein all except one factor or condition are held constant. By using this approach it has been discovered that many performance characteristics are determined by the ratio of the speed of the aircraft to that of sound. This ratio is called the Mach number after Ernest Mach, an Austrian physicist who did original research on shock waves. One important area of aerodynamic research is concerned with theoretical and experimental evaluation of performance in the transonic region near Mach 1.0—the speed of sound. As shown in Figure 6-1, the effect of speed on flight-control surfaces is reflected back to the pilot; through the transonic region, the abrupt changes in force required pose critical control problems. The supersonic transport was designed to cruise at Mach 2.7; new theoretical knowledge led to the compensated design with the improved characteristics shown. Research engineers are already working in the hypersonic region to explore the problems of flight or spacecraft reentry at speeds of up to 18,000 miles per hour (Mach 25).

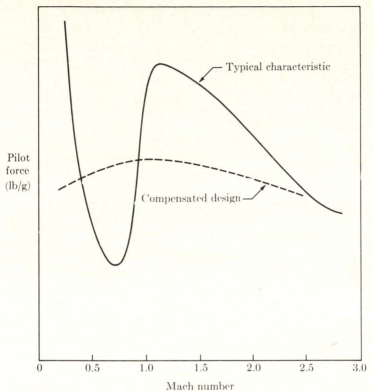

FIGURE 6-1
The effect of speed on the control force required of the pilot. *(The Boeing Company.)*

Chemical Engineering   What characteristics of a molecule determine its mechanical and thermal properties? How is lattice structure related to absorption and catalytic action? How are mass and energy transported within the reagents during a violent chemical reaction? How is the flame front propagated through a combustion chamber? These questions are typical of those asked by the research chemical engineer.

Note the long time span occupied by research in the history of the ultrahigh temperature film described in Figure 6-2. The value of research was recognized early in industrial chemistry and research engineers have contributed significantly to the mass production of synthetic fibers, plastics, food products, fuels, and pharmaceuticals by advancing our basic knowledge of chemical processes. Fundamental knowledge, experimental appara-

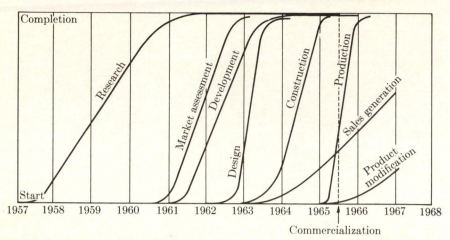

FIGURE 6-2
The overlapping activities involved in bringing a new product to the market. (*E.I. duPont de Nemours & Company.*)

tus, and mathematical tools have made it possible for imaginative men and women to create new products instead of waiting for their chance discovery.

Research Engineering

Research is now big business. Back in 1920 there were a hundred or so industrial research laboratories working on a total budget of less than $1 million per year. In the 1980s there were over 10,000 research laboratories with an annual expenditure of over $40 billion. Research is carried on in four major types of institutions: colleges and universities, industrial laboratories, organizations sponsored by the federal government, and research institutes. Over the past four decades, the government greatly expanded its emphasis on research and development and supports approximately one-half of all scientific research. However, industry has been increasing its support and now is an equal partner in support of research. In the past, federal funds were concentrated on basic and applied research related to defense, communications, and space exploration. In recent years there has been increasing emphasis on health care and industrial productivity whereas research and development on energy problems, pollution control, and transportation have been more cyclical and dependent on the economic, political, social, and international factors.

In most cases, a team of engineers and scientists is set up to tackle a given problem. Depending on the nature of the problem, the team will include experts in heat transfer, metallurgy, fluid flow, electronics, chemistry, physics,

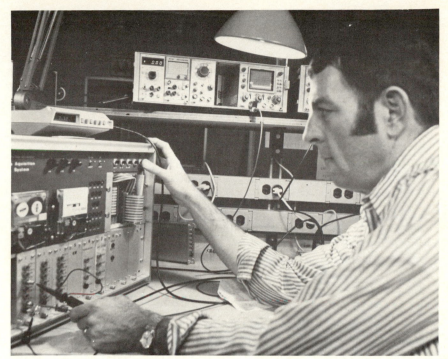

Electric vehicles have been around a long time, but electric autos that can compete with gasoline-powered cars are not available. A major obstacle to the widespread use of electric vehicles is the lack of a light-weight, high-capacity battery; a great deal of basic research is directed toward this goal. A better understanding of power-train characteristics is also needed. This research engineer is collecting data from thirty sensors connected to the complete power train of an experimental vehicle. The data will be fed into a computer model that may point the direction to optimum design.

or mathematics. Some persons will be strong in basic theory; others will be skillful in experimentation. In most cases, they will require specialized equipment and apparatus, computer facilities, and the assistance of skilled technologists. Although important research contributions will still be made by individuals, the day of the hungry experimenter working away in a cold garret is nearly over.

The actual research work is essentially mental, and results are usually in the form of ideas. These ideas may be expressed in the form of: a new law stating the relationship between physical factors, a graph presenting data on the performance of a new process, a description of a newly discovered phenomenon, an explanation of a previously observed phenomenon, proof of the accuracy of a proposed theory, or proof of the inadequacy of a theory. To be of value, the results must be communicated to the ultimate user. This may be done in a brief memorandum, a talk before a technical society, an article in a professional journal, or a formal report to the sponsor.

Qualifications for Research Work

Success in research usually requires a particular set of aptitudes developed by thorough training. What are the personal qualifications required for success in research? Since research requires doing what no one else has done, high intellectual ability is necessary. Because research involves work in an unknown field, imagination and acceptance of uncertainty are required. Patience and self-confidence will prevent discouragement over the many failures. Open-mindedness and intellectual honesty are important in any scientific work because without them the researcher may permit a preconceived idea or personal bias to obscure the truth. Ability to cooperate with others in a team effort is of extreme importance to the research worker, as well as the ability to express oneself to coworkers, assistants, and superiors.

What type of education is desirable? The research engineer must have a broad training in fundamental engineering sciences, along with the advanced specialized training that makes one valuable. In general, this combination is provided by a basic four-year engineering program followed by graduate work in a specialty. The basic course should include strong mathematics preparation and emphasis on engineering principles rather than practice. The graduate work should include advanced scientific concepts and an introduction to research work. The number of engineers completing the master's degree (one year) and the doctor's degree (three years) is steadily increasing. The seven years of intensive study culminating in an original contribution and a doctor's degree is particularly valuable in research.

DEVELOPMENT

From the research laboratories of the world there pours a continual stream of new knowledge. These new principles, processes, and products constitute the raw material for development engineers. Their job is to apply them to useful purposes. Knowledge about cryogenic behavior suggests a new memory element and an improved computer. A nonsticking plastic points the way to a new type of bearing. Data on supersonic airflow indicate control problems requiring new approaches. Synthesis of available knowledge plays an important role in development. When given the job of developing a new, better, or cheaper device or process, the development engineer first turns to the literature to see what already has been discovered by researchers in that field. He or she may follow through by talking directly with researchers, other developers, and designers at technical meetings or at on-site visitations. Not infrequently, sales engineers or marketing specialists may identify

In the studies of basic turbulent flame processes, important for developing
improved combustion systems, laser scattering can provide new information about
flame properties. The flame shown here is burning in a jet hydrogen-air combustor,
flowing from the center of the photo toward the bottom left. Two types of light-scat-
tering probes are used to determine the important flame properties. One type, called
Raman scattering (RS), is used to provide data concerning the molecular composition, density, and gas temperature. Another type, called laser velocimetry (LV), is used
to determine gas velocities. The intersection of the vertical RS probe beam passing
through the flame in the center of the photo with the two intersecting horizontal LV
laser beams defines a small test volume from which time-resolved data can be obtained by using a pulsed laser source. (*General Electric*)

an untapped need or a newly emerging product or service that will be the impetus for a new, challenging development project.

The development engineer occupies a strategic position between research and design. His or her achievement is usually the solution to a problem and frequently takes the form of a working model that is then designed for economic production. Although a given engineer may perform research, carry on development, and participate in design, especially in a small company, the three functions are quite distinct.

Example of Development Engineering

Suppose that marketing surveys indicate the need for a more convenient and more accurate method of measuring blood pressure. The necessary pressure sensors are available and the problem to be solved is how to calculate and present the measurements. The specifications are drawn up in consultation with physicians and then the project is turned over to a development engineer.

The engineer carefully reviews the specifications stipulating electrical, mechanical, and thermal characteristics. From her experience she identifies the probable bottlenecks or the criteria that are going to be difficult to meet. These might be extreme lightness, high accuracy, or low power consumption. A wide range of operator skills may also be needed. It is the combination of such specifications that constitutes the problem.

The next step is to review the literature to see how others have solved similar problems or how new research has pointed the way to possible solutions. She may have to translate, or have translated, articles by German, Russian, or Japanese experimenters. In a large organization the library staff works continuously at reading and summarizing articles of interest. Bibliographic computer searches of relevant data base may also be conducted using key words, names of leading authorities, or recently funded research projects or dissertations.

The next stage in development will be mental—attempting to create in the mind a new device with the desired characteristics. The emphasis here is on the bottlenecks; the engineer pays little attention to those portions of the circuit where a standard approach will suffice. She may lay out a circuit in block form, indicating only the functions to be performed. The new elements must be designed, and values must be calculated and specified.

A "breadboard" model may be constructed next. This is a spread-out, flexible unit which allows one to change circuit elements and to reposition units. Usually an electronics model is constructed by the engineer, but in other fields a technologist or a crew of skilled artisans will do the construction. This is not trial and error but rather a series of rational steps in which each change is based on measurement and observation. Versatile signal

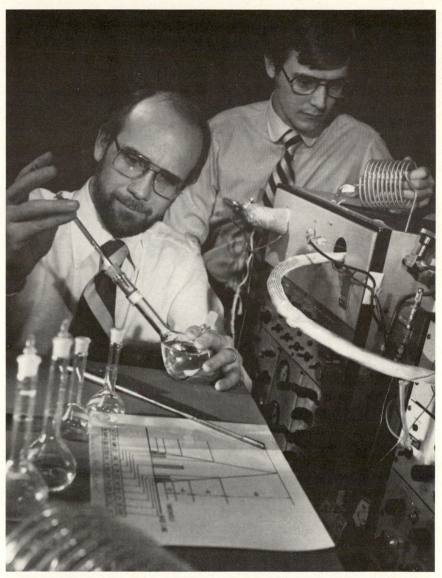

Trace amounts of organic compounds weighing as little as a few billionths of a gram have been automatically identified by a sophisticated analytical system invented in a research and development center. The system, which involves two gas chromatographs (right), a mass spectrometer, and "smart" electronics, provides analytical chemists with an important new tool for pinpointing infinitesimal amounts of organic solids, liquids, and gases in samples. The inventors prepare to inject a liquid sample into the system. This innovation promises to speed the identification of air and water pollutants, chemical contaminants in food and industrial products, cancer-causing substances, and certain regulatory hormones in human tissues. (*General Electric Company*)

generators, sensitive oscilloscopes, precise instruments, and calibrated circuit elements may be used to provide the necessary data.

On the basis of results obtained by using the breadboard model, a development model will be planned. This unit is constructed to meet mechanical specifications and is tested to see that it complies with electrical and thermal requirements. It may include components that were especially built for that job. Ordinarily, this preliminary model involves much handwork. When it is constructed, tested, and accepted by the sponsor, development is completed.

Invention

In the present organization of engineering, the inventor probably falls into the same classification as the development engineer. The original meaning of *invent* was "to come upon"—implying by chance. Many of Edison's inventions were the result of trying everything that could possibly produce the desired end. This method is still important in those fields of engineering that cannot be treated quantitatively.

One field in which invention is important is the development of fruit-processing equipment. In the past, fruits have had to be processed individually and by hand because of the variation in size and ripeness and the presence of imperfections. As a result, labor has been a big part of the total cost of the final product; mechanization can be justified in this type of situation. Included in this category are harvesters, peelers, pitters, and sizers.

Suppose an engineer is given the job of devising a machine to apply pesticides to plants without getting any of the poisons into the air or ground. First, he might make a patent search to check out all of the different methods of application. Ordinary spraying is not considered because it causes air contamination. Putting the new systemic insecticides into the ground for absorption by the roots is eliminated because irrigation water would carry the chemicals into streams. This leaves direct application of systemics to the plant as a possible approach.

Again, the second stage is mental—endeavoring to visualize various devices that would accomplish the desired result. The engineer could imagine a self-propelled vehicle carrying the operator, the insecticide, and the applicator. He could visualize brushes guided to plant stems by mechanical feelers substituting for human fingers. Perhaps photocells could substitute for human eyes. Laying out possible arrangements on paper would bring into focus other problems such as the feeding mechanism, power supply, and mechanical support.

The preliminary model would be bulky, but flexible. It would probably include some components originally designed for other uses and some parts

The development engineer applies the discoveries and results of research to some useful purpose. His efforts usually result in a working model of a circuit, or process, or machine that has desirable characteristics. Here engineers prepare a bench-scale reactor used to study ways of making coal liquefaction a more efficient process. They are adding a coal slurry to the reactor's preheater prior to an experiment aimed at identifying what happens during early stages of liquefaction. Coal liquefaction can be one answer to the energy problem if it can be made economically competitive. (*Sandia Laboratories*)

laboriously made by hand. Experimental operation of the model would reveal the model's shortcomings as well as its virtues. Perhaps several successive models would have to be built, each incorporating the improvements necessary to overcome the inadequacies of the previous model. This type of development engineering requires a strong background in engineering science with less emphasis on mathematical calculation.

Qualifications for Development Work

There are significant differences between the work of the development engineer and that of the research engineer, and these differences are reflected in the qualifications desired. It should be obvious that a successful development engineer must be ingenious and imaginative. His or her job is to create, and creativity is a highly marketable commodity. The development engineer should be skillful in experimentation, able to make accurate observations, and able to draw correct conclusions. In most organizations development is a group effort, and one must be able to cooperate with others, eager to make a contribution without worrying about getting credit for it.

Although ability to express oneself is important in all engineering functions, the ability to sell one's ideas is particularly important in development. The cost of conducting a development program may run into many thousands of dollars and involve shop facilities and additional people apart from the engineers. Management must be convinced with regard to the value of the idea and the ability of the one who proposes to develop it. On the other hand, a good development engineer must be willing to let go of his or her own "baby" and work enthusiastically on another's idea that looks more promising.

The technical education required for development varies with the field. In the more scientific branches such as astronautics, electronics, and thermodynamics, the broad training received in engineering science on the undergraduate level should be supplemented by special training on the graduate level. Employers recognize the value of advanced work in development by paying higher salaries to men and women holding master's and doctor's degrees. In the field of invention, creative skill and practical experience may be more important than academic training. In any event, development engineers are always working in new areas, and they need all the tools, techniques, and knowledge at their command.

DESIGN

In engineering work, the word "design" is used in two different ways. In general, design means to "create in the mind"; starting with a definite purpose, various elements are imagined to be combined so as to accomplish that purpose. In this sense a report is designed to persuade a certain type of reader toward a desired action. A manufacturing organization is designed to fabricate a product for a given market.

More specifically, design as an engineering function lies intermediate between development and production or construction. It is the responsibility of the design engineer to take the paper concept or working model from the development engineer and to prepare it for economic production in a factory if there are to be many or for construction on the site if there is to be only one.

Design for Production

Thorough knowledge of the properties of materials and the capabilities and limitations of fabrication processes is an essential part of the designer's education. Frequently a development model will contain elements that would not be suitable for mass production. The design engineer will figure out how to replace handmade elements with standard, and therefore inexpensive, items. In the model a part might be made of platinum to prevent corrosion, but in the production unit either the cause of the corrosion should be eliminated by proper design or a less expensive stainless steel employed. As a plan is prepared, the design engineer must have in mind how die cast parts might be used instead of individually machined items, and how industrial robots can accomplish such tasks as welding and painting.

Another important contribution by the design engineer is based on knowledge of the behavior of materials under the action of static and dynamic forces, electrical currents and voltages, and thermal stresses. Any factory laborer can select a timber beam large enough to pry up one end of a machine tool; if the first selection is too small and the beam breaks, a larger beam is simply chosen. The design of a structural member of an airplane costing $60 million and carrying 300 passengers at a speed of 1800 miles per hour at an altitude 50,000 feet above the earth is a different matter. Aircraft design has been refined to an extreme degree because of the need for safety on one hand and fuel efficiency on the other. Every ounce of structural material must be working at its maximum capacity, and to achieve this condition the design engineer must take advantage of every pertinent scientific advance.

Similarly, the electrical designer must specify the precise amount of copper that will carry the current without causing excessive heat that would shorten the life of a motor. In specifying the insulation to be placed around a steam pipe, the mechanical designer must calculate the cost of each additional inch of thickness and compare that cost with the corresponding saving in operation. As Wellington pointed out in his definition, the design engineer must do the job "well with one dollar. . . ."

In design for production, the engineer uses his understanding of the behavior of metals under the action of forces and thermal stresses, along with his knowledge of the capabilities and limitations of manufacturing processes. This automotive-type piston must make 60 trips per second the length of the cylinder and maintain a tight seal against exploding gases. It must be machined to very close tolerances, and the total manufacturing cost must be less than an hour's pay for a machinist. (*Educational Affairs Department, Ford Motor Company*)

In addition to working with the fabrication and performance of materials, the design engineer must come up with ideas for accomplishing such things as complex motion, heat transfer, lubrication, speed control, and operating safety. In some cases the desired result may be achieved mechanically, electrically, hydraulically, pneumatically, or electronically. The fact that there may be many solutions to the same problem usually comes as a shock to recent engineering graduates who have become accustomed to problems with only a single solution. After preliminary consideration the designer must decide on a single method and proceed. Making that decision is usually difficult for the inexperienced designer.

In general, the best design is the simplest. To paraphrase Wellington: "Anyone can plan a complicated machine; it takes an engineer to design a simple one." Another characteristic of a good design is that it is balanced. The power provided should not exceed that required, nor should one part of a device be of such great strength as to far outlast the other elements. The modern automobile is a good example of balanced design. At about the time the tires wear out with normal use, the battery dies, the brakes need relining, and a valve job is needed. For an extreme example of balanced design, read *The Wonderful One-Hoss Shay* by Oliver Wendell Holmes.

Everywhere we look we see engineering products. Just as a good architect feels responsible for creating a structure that contributes to the environment, the design engineer should strive to provide aesthetic as well as functional satisfaction. The supersonic airplane and the suspension bridge illustrate the principle that the best engineering design, in which each component functions well, is usually aesthetically satisfying.

Design for Construction

Many of the foregoing remarks are directly applicable to the design of structures. Structural designers must know the properties of steel, concrete, timber, aluminum, and masonry. While structural designers are concerned primarily with static loads, impact and other dynamic forces must be taken into account in providing for such factors as resistance to earthquakes or tornadoes. The number of structural elements is relatively smaller and includes beams, slabs, girders, columns, trusses, arches, shells, continuous frames, and diaphragms. Each of these has its particular application in creating a structure.

Certain aspects differentiate the design of a bridge, say, from that of an automobile. The structural engineer must have a great deal of information about the site—topographical information, soil and geological data, and information on stream flow, for example. Since only one bridge is to be built, there is usually no opportunity to modify the design and correct original errors.

The "bridge" from earth to space has as one abutment the huge Vehicle Assembly Building of Launch Complex 39 at the Kennedy Space Center. In just 12 months, design engineers from four cooperating firms provided for functions that were undreamed of a few years earlier. Every feature of the $100 million building, from the eight-tread Crawler-Transporter capable of carrying 16 million pounds at 1 mile per hour to the 456-foot-high doors through which space shuttles move on their way to the launch pad, represents an innovation. Using technical data gathered from earlier space flights, the engineer provides an easily transported and assembled vehicle ready for optimum performance in a hostile environment. Is this a job that would appeal to you?

Qualifications for Design Work

From this description of the duties of the design engineer, we can draw some conclusions regarding personal qualifications along with required training and experience. Design engineers must be creative because they are called upon to visualize mechanisms and structures that do not exist. They must be able to think on paper; the ideas for which they are paid become useful only when they are expressed in drawings and specifications. Design engineers must be able to decide which of the alternative solutions is preferable and proceed on that basis. They must be cooperative, since important designs are usually group efforts. They must remain open-minded and avoid becoming attached to an idea just because it is their own.

Although engineering design is not a closed profession, four years of engineering college is expected and at least one year of graduate study is desirable. The education must be broad because significant design projects involve several engineering disciplines and may include economic, political, and sociological factors as well as purely technical aspects. A good

The Vehicle Assembly Building is the world's largest building, enclosing 125 million cubic feet of air-conditioned space. This huge structure is designed to resist a once-in-a-hundred-years hurricane, and other structures nearer the launch pad are designed to withstand a 1.2-kiloton bomb blast that could result from an accidental explosion. The space shuttle is assembled indoors and moved out through doors large enough to admit the United Nations building on the Mobile Launcher Platform. Each of the three components of the space shuttle requires engineering design varying from microprocessor circuits to some of the largest systems ever created. (*NASA*)

design is economical in terms of materials, fabrication, installation, operation, and maintenance. The designer needs the flexibility of mind that is developed by familiarity with the techniques of analysis used in a variety of fields. Since design is an iterative process of synthesis and analysis, the computer is a valuable aid. Designers should be able to take advantage of the computer by allowing it to enhance their analytical ability and to perform much of the labor of data reduction and graphical presentation. In addition, they must be able to deal with aesthetic considerations, and frequently they must be guided by the physiological limitations of the user. A knowledge of law makes it easier to understand contracts and to check on patents.

CONSTRUCTION

The construction industry constitutes a large segment of the national economy and accounts for a significant proportion of the gross national product. Within the industry are found engineers performing each of the eight major functions. In this discussion, however, the term *construction engineering* will be used to mean the work of building structures that have been designed previously and that will be operated subsequently.

The construction engineer receives from the design engineer a set of plans and specifications. These vary in complexity from a single sheet with sketches and notes to a set of hundreds of drawings and thousands of pages of specifications. It is the job of the construction engineer to turn the visions of the designer into reality—to translate pencil lines into steel and concrete.

Heavy Construction

The majority of construction engineers are engaged in heavy construction—buildings, highways, bridges, sewage and water systems. A secondary field involves the erection of refinery units, towers, and tanks, and the installation of air-conditioning systems, boilers, and power-generating equipment.

From the time plans are first received for bidding until the dedication ceremonies are held, the construction engineer is active.

Estimating and Bidding   Ability to predict cost is one of the distinguishing characteristics of the engineer. Nowhere is this more true than in construction. When plans for a large job are received, the construction company's top engineers are called into conference. The project as a whole is discussed, and then it may be broken down into parts for further consideration by specialists.

The first step is to plan the most economical construction procedure. On a hydroelectric project, for example, the questions to be answered might include the following: What structure will be required to reroute the stream during construction? How much soil must be removed before a suitable bedding is reached? What type of machinery will do this job best under the circumstances? What is the best location for the concrete plant? What is the best source of sand and rock? What money-saving methods can be employed in placing the concrete, in constructing forms, and in removing forms? What facilities will be needed to house and feed the workers, to provide recreation, and to repair mechanical equipment?

Simultaneously with formulating the plan of construction, estimators are performing *quantity takeoff*. This consists of calculating and estimating the quantities of excavation, formwork, concrete, reinforcing, and piping. Each company keeps careful records to provide a basis for estimating unit costs, for example, the cost of placing one yard of concrete or fabricating one square foot of formwork under specified conditions. Since each job is dif-

In planning the construction of a bridge, the engineer must answer such questions as: How much excavation for foundations will be necessary? What is the best location for the concrete plant? How should the forms be erected and removed? For this bridge on the Caracas-La Guaira Autopista in Venezuela, the form for the central arch was designed, built on the riverbed, hoisted into place far above, filled with concrete and prestressed steel, and lowered. (*Courtesy Civil Engineering*)

ferent, these are estimates only, and a great deal of judgment based on experience is required.

After determining the plan of construction and the quantities involved, and after estimating the cost of each operation, the company is prepared to make a bid for the job. Usually a large part of the work will be performed by subcontractors. The bids for the electrical and mechanical work, and so forth, must be carefully checked to ensure their adequacy, and then incorporated into the general bid. This is a critical moment in the project, and the office is the scene of frantic checking, rechecking, and last-minute changes. If the bid is too high, all this work is for naught. If the bid is too low the company may face financial ruin. In general, several companies are pitting their construction skill and financial judgment against each other in bidding for the contract.

Fieldwork   Once the contract has been signed, a project organization is set up and work begins. In general there will be fieldwork and office work. Fieldwork starts with a surveying party sent to the site to obtain additional data and lay out the project as designed. The survey crew sets grades, determines cut and fill, and locates critical points such as boltholes for major pieces of equipment.

Construction of access roads and auxiliary buildings begins immediately. Usually some excavation is necessary, and that is followed by erection of forms and other falsework. The construction engineer must take advantage of labor-saving machinery and timesaving methods. Modern construction makes use of gasoline-, diesel-, and electric-powered trucks, tractors, bulldozers, shovels, cranes, cableways, and pumps. Helicopters are used to place steeples on churches and transmission towers in wilderness areas. Prefabrication, prestressing, tilt-up, and lift-slab methods offer special advantages. The construction engineer advises the designer regarding modifications that will facilitate construction or improve the structure without adversely affecting its operation.

In addition to working with materials, machines, and methods, a primary responsibility of the construction engineer is people. Obviously every project is a team effort, and the selection of capable supervisors is extremely important. The safety and welfare of the individual workers must be ensured. Good relations with the unions representing the various crafts must be maintained.

Office Work   The office engineer drafts, calculates, and accounts. The plans frequently need interpretation and clarification, and auxiliary drawings may be prepared. Also, "as-constructed" plans are prepared where the finished structure differs from the original drawing. Progress reports indicate the rate of completion of various portions of the project and form the basis for partial payments to subcontractors. In addition to having a detailed knowledge of the plans and specifications, the engineer must be familiar with construction procedures and costs.

Project managers supervise the work of the construction engineers, but their primary responsibilities are of a nontechnical nature. They troubleshoot on a high administrative level, solve right-of-way problems, and maintain good public relations. They keep an eye on the subcontractors and try to prevent difficulties from getting to the lawsuit stage.

Mechanical Construction

The overall problems involved in the erection of a refinery unit or the installation of an air-conditioning system are similar to those in heavy construction. There is the same emphasis on careful planning or procedure, development of timesaving methods, use of laborsaving machines, and on efficient teamwork among the various craftworkers. Also similar are the estimating and bidding procedures.

The major difference is that this type of construction involves more placing and assembling of complete units that have been fabricated elsewhere. Piping, ductwork, wiring, and controls are important, and the engineer on this type of job needs a strong background in electrical and mechanical practice. Some large organizations combine design and construction functions and operate under the title of engineering contractor. Actually, however, the personal qualifications and training required are quite different for the two functions.

Research, development, and design engineers work in idealized situations or on paper. Construction engineers are concerned with the real thing. They are continually working under pressure, and overtime is accepted practice. They must get the work done with what is available; they must take risks, make decisions, go ahead, and learn by mistakes. They must be ready to move anywhere, anytime, and live under all sorts of conditions. Their most difficult problems require a technical background, but their ability to work with people determines their success or failure.

A basic engineering-degree program will provide the necessary knowledge of materials, forces, structures, and electrical and mechanical equipment. Skill in writing reports is also important. Background in business law, economics, psychology, and labor relations is desirable. Training in construction methods and equipment may be available in college; it may have to be gained on the job. In any event, construction engineering requires experience and judgment that can be obtained only through actual practice. Summer positions as surveyors, inspectors, or timekeepers provide young construction engineers with an opportunity to gain this essential training.

PRODUCTION

The price at which an article can be sold is fixed by competition in the open market. At the other end, the cost of raw materials and labor is fixed. Whether or not a company can operate at a profit depends on its ability to perform the conversion of raw materials to finished product more efficiently than its competitors. This conversion is the responsibility of the production engineer.

The work of the production engineer is quite similar to that of the construction engineer. In the usual company organization, production is directly under management rather than under engineering. The emphasis is on *how* to accomplish after the design engineer has specified *what* to accomplish. The production engineer advises design engineers on designing for production, selects tools and processes, and schedules production.

Advising Design

Although it is essential for design engineers to be familiar with the capabilities and limitations of fabrication processes, they cannot afford to take time to become expert. Production engineers advise on materials, processes, and

The production engineer must integrate the flow of raw materials, subassemblies, and completed units with market demand. Here a production engineer explains a detailed drawing of various stages as the parts move down the engine assembly lines. (*John Deere Tractor Works*)

procedures. If a particular material is in short supply, they suggest an alternative. If the proposed corrosion-resistant material is difficult to fabricate in the specified manner, they may suggest a change in material or a change in process. Production engineers must be able to predict the effect on cost and time of reducing the allowable tolerance from ± 0.002 to ± 0.001 inches, for example. Frequently, a small change in design will greatly facilitate assembly and therefore reduce cost.

Process Planning

In the case of a mass-production item, a great deal of attention is given to selecting the fabrication process, the major fabrication steps, and the sequence to be followed. Each process has its advantages and limitations. Sand casting is good for mass production but expensive for a few units and results in a rough finish. Die casting gives precise results and smooth finishes but is usually limited to small items and relatively weak materials. Light sheet metal can be formed rapidly by presses or brakes; heavier sheets can be cut and welded, but this is a relatively slow process and more adaptable to small-quantity production. Machine tools such as the lathe, milling machine, and shaper are capable of precise work but require a skilled operator; output is relatively slow. Automatic assembly lines and industrial robots increase the output, but are prohibitively expensive except for mass production. Tape-controlled machines offer the flexibility of general purpose tools with the speed and precision of automatic machines.

The production engineer must decide on the individual processes and the combinations of processes in fabricating each component, always working within the limitations of the equipment on hand or prepared to justify the purchase of a new machine. The overall production sequence must take into consideration the routing of raw materials and subassemblies. In every decision the production engineer must seek the specified result at the lowest possible cost.

Tool Engineering

One characteristic that distinguishes humans from animals is the humans' use of tools. Sometimes production engineering is referred to as tool engineering, but more properly tool engineering is that phase of production that has to do with major and accessory tools. The major machine tools represent investments of many thousands of dollars; even a small precision lathe and accessories may cost as much as a six-room house. To recover the investments on a given machine requires efficient utilization—utilization at a maximum rate and for those processes that it does best. The production engineer makes replacement studies to determine the point in the life of a machine at which it should be replaced by a new or improved tool.

One important phase of production is tool design. Cutting and forming tools are usually designed and built on the job, whereas the major tools are usually purchased from outside sources. Dies are required for stamping, forming, drawing, and extruding. Cutters for lathes, mills, and shapers may make use of the new, long-lasting carbides in order to reduce costs of high-speed production. Special fixtures are designed to hold work in position and jigs are employed to guide the tools, thus cutting down time and permitting the use of less skilled operators.

Production Scheduling

The production department must work closely with the sales and purchasing departments. The flow of raw materials, parts, subassemblies, and completed units must be integrated with market demand. In the automotive industry this scheduling becomes extremely complex, and computers handle the mass of data. Your purchase of a new car of a certain model and color and your choice of accessories may be recorded in the form of a punched card. Feeding the card into a computer in Texas will automatically report the necessary orders to an assembly line in Detroit, a purchaser in New York, and a supplier in Akron.

The production engineer must organize, update, and manage these computer scheduling systems that provide for inventory control, order dispatching, equipment scheduling, optimum machine loading, and quality control. If not well planned, these operations may cost more than the actual production itself. Frequently, the production engineer must break bottlenecks. Sales asks for more output, maintenance demands more time for repairs, purchasing reports that certain materials are unobtainable, management objects to so much overtime, and, just then, engineering insists on making some design changes. It is under circumstances like these that production engineers earn their salaries.

Qualifications for Production Work

Since the functions of production and construction are much alike, the desirable qualifications are quite similar. The production engineer works under constant pressure and deals with machines and systems that are pushed to the limit. A feeling for what machines and people can do is important. This is a good job for a planner, but it is no place for a worrier. Roaming throughout the plant involves more physical activity than design. Ability to cooperate with subordinates and representatives of other departments is essential. The ability to sell an idea to management is extremely important when that idea involves the investment of millions of dollars.

The technical problems faced by the modern production engineer cover practically all phases of engineering, so strong training in the basic engineering sciences is required. In addition, specialized training is needed in production processes, metallurgy and heat treatment, engineering economy, and quality control. Practical training may be obtained in college laboratories, on visits to plants, or in vacation work. Starting assignments are frequently in the shop where production operations may be observed.

OPERATIONS

After a piece of equipment has been developed, designed, and produced, it still must be operated. Any modern 4-year-old can operate a sophisticated television set, and most 16-year-olds are licensed to operate a 200-horsepower automobile. However, the operation of a computer center, launch facility, or nuclear power plant requires engineering training. In a manufacturing operation, the supervision of the manufacturing facilities—buildings, equipment, and utilities—is called *plant engineering.* The control of public utilities or processing plants in order to obtain maximum reliability and economy is called *operation engineering.* In both activities the maintenance of plants and equipment is of sufficient importance to justify special discussion.

Plant Engineering

Suppose that you, a plant engineer, are awakened some early morning by a call for help. "Get over here as fast as you can. Unit 5 has gone off production and it's costing us $1000 an hour." The first question is, What is wrong: Is it a failure in power or controls, or has a major component of the unit failed? Once the difficulty has been located, the next question is: Should the defective element be repaired, or should it be replaced? The answer to this question must be on a dollars-and-cents basis, taking into account a great many economic factors. Once the unit is back in operation, the question is: How could proper inspection and routine maintenance have prevented this costly breakdown?

The plant engineering department is a staff organization whose function is to provide service to the producing departments. In general, efficiency and dependability of operation are the principal objectives, but they must be expressed in terms of economics. In some types of processing plants, for example bakeries and glass plants, efficiency must be sacrificed in order to secure dependability and to avoid service interruptions.

Plant engineers plan new facilities, supervise their construction, select and install equipment, and supervise the operation and maintenance of the facilities. They are usually responsible for plant safety, including safeguards for the individual workers and provisions for fire prevention and fire fighting.

A quality control engineer working on the production line is performing a time study on a tractor assembly line. Well planned assembly and inspection procedures and carefully selected test equipment will minimize production difficulties and maximize product quality. (*Deere and Company*)

The complete crew includes electricians, plumbers, pipefitters, millwrights, instrument repairers, and other craftworkers and technicians.

Operations Engineering

A small percentage increase in the amount of high-octane gasoline obtained from crude oil in a cracking plant may be worth $1 million over the course of a year. A slight reduction in the amount of fuel oil burned for each kilowatt-hour of electrical energy produced in a steam power plant may represent the difference between profit and loss. Consequently, many of the engineers employed by power companies, refineries, waterworks, or telephone companies are primarily engaged in operation.

In a nuclear power plant the operation engineer's duties might include keeping in close touch with daily operation, analyzing actual operating data and comparing them with design values, notifying design engineers about troublesome features and recommending desirable improvements, monitoring radiation levels and safety devices, gathering data for special studies and recommendations, investigating malfunction reports and providing liaison with equipment manufacturers, supervising installation of new units or con-

struction of new plants, and setting up inspection and maintenance procedures.

Maintenance Engineering

It recently was estimated that for every industrial worker there is $40,000 invested in machines and equipment. Obviously no organization can afford to let such an investment go unprotected. While the actual maintenance will be in the hands of skilled craftworkers and technicians, the planning and supervision require engineering. Maintenance is always a responsibility of operation and in many cases is supervised by the plant engineer or the operating department.

Aeronautical maintenance is a highly developed engineering activity of special interest. In most industries, breakdowns are costly and should be minimized; in the air transportation industry, however, the failure of a single part may lead to a catastrophe. Elaborate procedures have been devised to ensure that critical parts are replaced before failure. During operation, tests are performed to check the overall condition of components. Complete overhauls are scheduled after a specified number of hours of operation. In an engine overhaul, the engine is completely disassembled and every part is inspected. Certain parts are replaced regardless of apparent condition. Other parts are examined using equipment that reveals cracks invisible to the naked eye. X-ray inspection is used on castings and weldments to make sure there are no internal imperfections that could cause failure.

In all operation and maintenance work, the keeping of complete, accurate records is essential. Frequent analysis of records of performance or failure permits modifying the operating conditions for greater reliability or efficiency. On the basis of the data, design changes are recommended. A large aircraft maintenance base will include an engineering department that provides design and drafting service for the operating and maintenance departments.

Teamwork between equipment manufacturer and utility companies has resulted in a computer-based system that could cut hours off the time required to restore electric service in the event of a power failure. The power restoration system quickly locates a trouble spot (such as a downed power distribution line) and automatically reroutes electric service around it by remotely opening and closing various switches. Instead of a whole neighborhood being without power while repairs are made, only the few homes in the immediate area of the fault would be affected. Here, an operations engineer examines the microcomputer circuit board that serves as the "brains" of a remote unit that opens or closes a pole-top switch on command from the system's central computer. (*General Electric*)

The automated production line, such as this unit for producing fine dinnerware, replaces human hands in difficult, dangerous, or monotonous activities. Maintaining efficiency in such complex equipment without costly breakdowns or shutdowns requires careful planning based on a thorough knowledge of forming and firing processes along with an understanding of hydraulic, mechanical, and electrical operations. The maintenance engineer consults with the designer on features that insure trouble-free operation. (*Corning Glass Works*)

Qualifications for Operations and Maintenance Engineering

The operation engineer is more concerned with the performance of actual machines and systems than are the engineers engaged in other functions. This emphasis is reflected in the characteristics of the successful engineer in this activity.

This function is for men and women who like machines and equipment and enjoy trying to get the most out of them. They should be able to track difficult problems and work under pressure. They must be able to get a lot out of people as well as machines. They supply the leadership and the brainwork for a crew of craftworkers or technicians who have the know-how and manual skills. They should be methodical, able to plan well, skillful in gathering and analyzing operating data, and able to translate technical situations in terms of economics. Because they perform a service function, they must be able to cooperate with other departments of the organization. Ability to work with abstract scientific or mathematical concepts is of minor importance.

As with most functions, working in operations and maintenance covers all the basic fields of engineering—civil, electrical, mechanical, chemical, metallurgical. Experience with actual machines, processes, and controls may be obtained in college laboratories but must be supplemented by on-the-job training. College work that includes a background in economics and statistics is valuable for obtaining many positions, as is summer work in some of the skilled crafts.

SALES AND APPLICATION

One of the duties of operation engineers is to select equipment and machines. For much of their technical information, they rely upon the sales engineers representing the various manufacturers. Sales is the vital link between the production of a technical item and the operation wherein it becomes useful in the form of engineering equipment. The day of the glib salesperson whose major attributes were a smooth tongue and a large entertainment fund is over; in the forefront today are engineering graduates who know the technical capabilities and limitations of their products and who, through that knowledge, are able to improve their customers' operations. Where major pieces of equipment are involved, sales engineers are supported by technical specialists called *application engineers* and field representatives called *service engineers*.

Sales Engineering

The problems of the sales engineer usually arise as a result of a customer's need. For example, the representative of an electrical manufacturer may receive a call as follows: "We've just placed another processing line in operation and our main transformers are overheating. What can we do about it?" The sales engineer visits the plant, studies the equipment being operated, and examines the monthly meter readings. Frequently, in a situation like this, the overheating results from what is called "low power factor" caused by some equipment drawing a relatively high current for a given power. Instead of replacing the transformers, the sales engineer can show the economic justification of placing electrical capacitors on the line thus improving the power factor, eliminating transformer overheating, and probably reducing the monthly energy bill.

This example is typical of many in which the sales engineer must first solve the technical problem, then specifiy the equipment that will accomplish the desired result, and finally sell the customer on the economic desirability of the solution he presents. The technical work may vary from merely recommending the proper size for a circuit breaker to planning an elaborate processing unit complete with power supply, major units, accessories, controls, and instruments. In many respects the sales engineer performs a function similar to that of the architect. Starting from a stated purpose he must combine elements to accomplish that purpose efficiently, safely, and reliably. In the broader sense, this is design. Wherever possible, of course, a sales engineer uses products from his own company, but the customer's good will is so valuable that he must avoid proposing an inferior or unduly expensive unit. Frequently, several sales engineering representatives will submit bids on the same job, and competitors will be quick to point out any weaknesses in a proposal.

The daily work of the sales engineer may include the following: studying new products by means of company literature and operating instructions; traveling between the customers' plants, the home office, and various factories; planning for "creative selling" where a customer may be unaware of need; analyzing customer complaints; conferring with company associates and attending sales meetings; representing the company at technical meetings and entertaining important customers; reporting field problems and innovations by competitors to the design engineers; and reporting on sales progress.

Sales engineers are unique in two respects. First, they personally represent the company in many contacts and transactions. They are directly connected with company income and, knowing the market and competition, they influence design and production. Because of their background, sales engineers can step up into management with greater ease than can specialists in the development laboratory or the design office. Second, sales engineers deal with a wide range of persons—research scientists needing unusual equipment, process engineers with rigid technical specifications, practical persons interested only in overall operation and maintenance, and city council members interested primarily in cost or environmental factors. Sales engineers must be able to speak with each of these on the appropriate technical level.

Application Engineering

When the products are more complex and involve more highly technical features, there is a tendency for the sales engineer to concentrate on customer contact and turn over the design problems to application specialists. These application engineers work in a centralized office and are available as consultants to sales engineers and customers in that market area.

One large electrical manufacturer maintains offices in major cities that are staffed by specialists in industrial power, chemical and petroleum

Seldom does a sales engineer demonstrate his new product for such a large audience. In most cases, however, he does have the same necessity to adapt his presentation to the interest of his customers. These farmers are interested in the performance and cost of this new gas-turbine-powered, hydrostatic-driven tractor, and a demonstration in the field is most effective. Back at the factory, design engineers will be eager for reports by the sales engineer on maintenance problems and innovations by competitors. (*International Harvester Company*)

processing, central-station equipment, communications, or transportation applications. In addition to knowing the standard products, these engineers are qualified to modify existing designs or design special apparatus to meet special needs. For example, for a petroleum processing unit, it was necessary to provide an elaborate computer controller that would analyze the crude input and adjust the operating conditions to optimize the output of the principal product. The resulting design was so successful that it was patented and made a part of the regular product line.

Service Engineering

Obviously an elaborate unit of this type cannot be loaded on a freight car, sent off, and forgotten. The performance of the unit and the customers' satisfaction are too dependent upon proper installation and operation. Experience has proved the desirability of including in the contract responsibility for installation, field testing, and operator training.

As soon as the equipment is delivered, the field or service engineer gets on the job. She may use her own crew, the contractor's people, or a crew she recruits for the installation. She may have to plan foundations, utility installations, unloading facilities, and erecting methods. After the unit is assembled, she checks each component, making any necessary adjustments and correcting any travel damage. With the assistance of the customer's operators she starts up the unit and checks its performance. Usually she can expect a series of difficulties, and her ability as a troubleshooter is important. She is also responsible for training the customer's crews in proper maintenance procedures.

Qualifications for Sales Engineering

While sales, application, and service functions may differ in their responsibilities in a large company, the personal qualifications and training required for all three are quite similar; in many cases one individual performs all three activities. The distinguishing characteristic of the so-called sales personality is the ability to establish communication with strangers quickly. Sales engineers must be friendly, easy to get acquainted with, and good listeners. They should inspire confidence in their knowledge of their product as well as in their business integrity. They should be courteous and tactful and undismayed by indifference, resistance, or insult. Since they represent their company, appearance is important. They must be able to interpret the customer's requirements and to express their own ideas effectively. They should have good business sense. They should not mind traveling.

Sales engineers possess this type of personality, and, in addition, the technical ability to solve a wide range of engineering problems including design, operation, and maintenance. They must be ingenious because they are called upon to meet novel situations. They must make field decisions on important matters and run the risk of making mistakes. They must be able to deal with persons on all technical levels.

Sales engineers can expect to work on problems in all fields of engineering, and therefore they need broad training. Because much of their work is on a highly technical level, an engineering degree is almost mandatory. Their training should include the fundamentals of equipment, processes, and controls. Training in business law, accounting, and economics is also valuable.

College training does not lead directly to sales. Usually a sales trainee spends a year or so in manufacturing or service departments to become familiar with the products and their characteristics and to develop the necessary judgment. Frequently the first sales assignment will be in conjunction with a more experienced engineer.

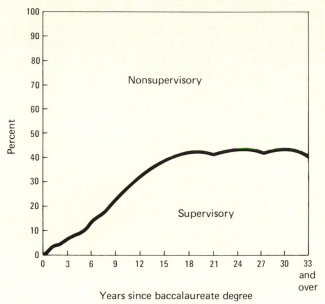

FIGURE 6-3
Percentage of engineers in supervisory positions. *(Engineer's Salaries, Special Industry Report, Engineering Manpower Commission of the American Association of Engineering Societies, Inc., September 1981, New York, pp. 27–28.)*

MANAGEMENT

Surveys of engineers and engineering graduates consistently indicate that within twenty-five years after graduation almost half of engineering graduates are performing managerial, supervisory, or administrative functions. Figure 6-3 provides data from the 1981 salary survey of the Engineering Manpower Survey indicating the increasing percent of engineers who became supervisors with increasing experience.

The increasing participation of technically trained persons in top management is revealed in studies of the educational background of the chief executives of the largest U.S. industrial enterprises. A study by *Scientific American* indicated that half of the top management positions in major U.S. industries will soon be held by persons trained as engineers or scientists.

The pioneering work of Mabel Newcomer, a study by Market Statistics Inc., and research by the Harvard Graduate School of Public Administration moreover indicates a continuing rise in the percentage of top industry officials who were trained in science or engineering. After a steady rise of less than 10 percent in 1900 to 20 percent in 1950, the fraction of executives with technical backgrounds climbed to 40 percent in 1960 and to 50 percent in 1980. These earlier studies as well as the more recent studies of the younger executives—those in the 35 to 45 age group—indicate that the majority have been trained as engineers or scientists. If technically trained persons are promoted in proportion to their numbers, it is expected by the turn of the century most of the top industrial managers will have backgrounds in engineering or science.[1]

What is management and why are engineers so desirable in management positions? One way of illustrating the relationship of management to the other engineering functions is as follows: Research and development determine *if* a thing can be done, design and production determine *how* a thing can be done best, and management decides *whether* to do it at all. Management must decide how the assets and capabilities of an organization are to be used. The abilities that characterize a good engineer in other functions are useful in making these decisions.

Purposes, Policies, and Decisions

Management has a threefold responsibility. An obvious responsibility is to the owners: the proprietor, the partners, the stockholders, or government or regulatory agency which may represent the public interest or national concerns and priorities, depending on the type of organization. It is now generally accepted that management is responsible also to the workers, and no longer is labor regarded as a commodity to be purchased at the lowest price possible. The third major responsibility is to the general public which must benefit from management's operations.

In line with these responsibilities, management must set up the purposes of the enterprise, establish the policies by which these purposes are to be achieved, and make decisions consistent with these policies. Some industrial companies are organized to perform all engineering functions from research to sales, while others concentrate on a narrow range of functions. Some specialize in a particular type of product, while others endeavor to spread their efforts over a wide variety.

Money

The assets of an enterprise include *money, people,* and *things* such as plants, processes, materials, methods, and machines. The responsibility to the

---

[1] "U.S. Industry: Under New Management," *Scientific American,* 1963; and *Manpower Comments,* 1980, vol. 17, no. 9, November 1980, pp. 28–29; vol. 17, no. 10, December 1980, p. 7.

owners is primarily in terms of efficient money use. Each year billions of dollars are spent for new plants, additions to existing plants, and modernization of old plants. Capital is available for these purposes in the expectation of earning a profit. Having determined to expand into a new operation, say, management must select a means of financing. This may be done from previous savings, current earnings, or some outside source.

The decisions by management are usually based on data supplied by others. Some questions asked are as follows: Shall we initiate a research program in this field? Is it justifiable to sponsor a project to develop a new use for this material? Is it time to redesign our major product? Can we afford to expand our production facilities in expectation of greater demand? Shall we undertake an expensive advertising campaign? In each case the management group depends on the various departments for specific information concerning the advantages and disadvantages of a proposal. Usually the information is embodied in a written report accompanied by charts and graphs and, probably, an oral presentation. The ability to write reports and present ideas is of great importance to all creative persons, especially engineers.

People

The work of the enterprise and the ideas that result in progress and profit are contributed by individuals. Management's part consists in selecting, training, and directing people. A well-defined organization plan lists the specialists involved, specifies their duties and responsibilities, indicates their relationships to others, and establishes the authority by which they function. Men and women are assigned to particular positions on the basis of their abilities or, occasionally, positions are modified in scope to fit the individual. People must be trained to increase their effectiveness in current positions and prepared for greater responsibility in future positions. Management always has the duty of directing the abilities and energies of its workforce toward the established objectives of the organization.

Things

In a manufacturing organization, for example, most of the decisions involving technical aspects of the operation are made by technical people. Management is concerned with broader problems involving large amounts of money or numbers of personnel. The adoption of a new major process, the location of a plant, or the replacement of expensive machinery is usually decided at the management level.

Plants, structures, and machines seldom live out their physical life. Long before a machine is worn out, it usually has become economically unusable. The development of better materials, improved processes, and more efficient machines renders old machines obsolete. When should a machine be replaced by a new, more efficient unit? The answer to this question depends on the events of the future and cannot be stated exactly. The practical answer must be based on judgment developed in the past and prediction for the future. Management policy may be that the cost of a capital investment must be recovered through savings in operating expenses over a five-year period. Once that decision has been made, engineers can determine if and when replacement is justified.

Qualifications for Work in Management

Management is generally recognized as a possible goal for engineers in other functions. The requirements for success in management are indicated by the emphasis on ability to solve problems involving people and money as well as things.

A manager is basically a leader with good business sense. Young men and women who have been selected by their classmates for school and club offices and have been successful in carrying out the responsibilities of those offices have already demonstrated leadership qualities. Management requires a broad viewpoint, and frequently the manager type of person is less interested in technical specialization than is the design or research type. He or she should be interested in social and economic factors as well as scientific factors and willing to look to the future as well as the present. Most successful engineers have strong motivations that carry them to the top. For the development engineer the desire for credit and recognition may be the motivating force; the management engineer is usually more interested in position and power. One important characteristic of the manager is sensitivity in human relations, an understanding of the basic desires for security and recognition, and acceptance of the responsibility for contributing to the personal welfare of his or her employees.

The usual route to management starts with technical positions, and proceeds with supervisory jobs and administrative responsibilities. Since it is through their first technical assignments that young engineers establish themselves, strong engineering training is essential. Some engineering functions provide a broader experience and, therefore, are more likely to result in opportunities for entering management. Production is a primary function that affects profit directly and involves work with people and money; sales engineers represent the company to the customer and are familiar with the market and competition. In a science-oriented industry the development engineer plays a key role in the generation of new products. Many of the industrial managers of the future will come from these engineering functions.

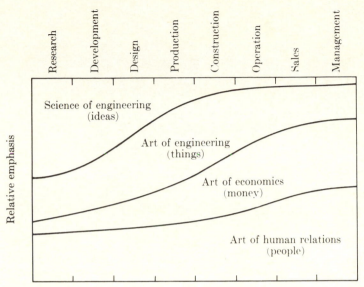

FIGURE 6-4
Activity chart for engineers in different functions.

THE ENGINEERING SPECTRUM

In general, engineers work with *ideas* (abstract scientific concepts and principles), *things* (machines, materials, structures, circuits, taxes), *people* (employees, associates, supervisors, customers), and *money* (financing, costs, taxes, prices, profits). The relative extent to which engineers use knowledge in these four areas is indicated roughly in Figure 6-4.

The variation in emphasis among the functions of engineering illustrates the fact that engineering is not a single activity but rather a spectrum of activities offering a variety of careers. This functional classification of engineering is particularly helpful for the beginning student in engineering. Right now you should begin thinking about your own aptitudes and personal qualifications and comparing them with the requirements for the various functions. Early in your engineering career you should decide whether you prefer to work with ideas, things, people, or money, and allow this decision to guide you in your selection of optional courses and help you get the most out of your engineering education.

Rather than restrict your choice to a single function at the present time, you may wish to think in terms of broad areas. Area *A* in Figure 6-5 represents the highly scientific functions of research and development. What are

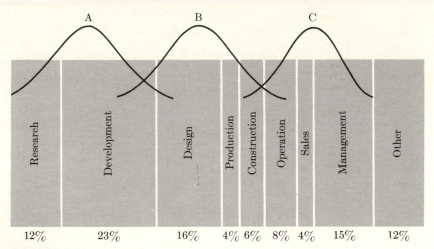

FIGURE 6.5
Distribution of engineers by function. *(National Career Development Project, Purdue University)*

the distinguishing requirements for work in this area? Area *B,* which includes design, construction, and production, emphasizes more practical problems. What qualifications should you have for this area of work? Area *C* involves personal relations and contact with people in operation, sales, and management. Does your past experience indicate that you would be successful here?

Branch, Function, and Rank

Interest is usually associated with the branch of engineering while aptitude and training should determine the preferable function. Ultimately, defining your choice of career will require two selections; for example, you may decide on electrical-development, chemical-operation, mechanical-design or petroleum-sales.

Tables 6-1 and 2 reveal how the function performed is related to degree level, experience, and branch of engineering. Engineers at the bachelor's degree level are likely to be in design and operations; those holding the doctor's degree are most likely to be in research and teaching. Inexperienced engineers are likely to be in design; those with many years of experience are most likely to be in management. Aeronautical, agricultural, and chemical engineers are likely to be in research, development, and design;

TABLE 6-1
FUNCTIONS OF ENGINEERS BY DEGREE LEVEL AND EXPERIENCE

Function	Total	Degree level			Years of experience		
		B.S.	M.S.	Ph.D.	1–5	6–10	11+
Research	9%	4%	16%	29%	9%	8%	7%
Development	11	8	18	9	12	12	10
Design	20	25	19	3	29	16	7
Production	8	10	6	2	10	8	6
Construction	4	6	1	0	5	2	1
Operation	8	10	5	1	9	8	5
Sales	3	4	1	0	3	3	3
Management	19	18	15	22	8	25	37
Consulting	6	5	9	8	4	7	11
Teaching	3	1	2	25	1	2	7
Other	9	9	8	1	10	9	5
Base numbers	2250	1158	560	98	1148	616	451

Source: National Career Development Project, Purdue University, W. Lafayette, Ind., 1982.

electrical and mechanical engineers are most likely to be in design; civil engineers are most likely to be in construction.

Titles of practicing engineers must also indicate their rank or the degree of supervision they exercise. Typical rankings are:

TABLE 6-2
FUNCTIONS OF ENGINEERS BY BRANCH OF ENGINEERING

Function	Total	AAE	AgE	ChE	CE	EE	IE	ME	MiE	NE
Research	9%	18%	29%	14%	4%	6%	3%	8%	19%	6%
Development	11	17	4	19	3	22	5	14	11	14
Design	20	16	35	24	33	44	1	34	9	19
Production	8	4	9	9	4	8	18	14	8	7
Construction	4	0	1	1	17	2	0	4	0	6
Operation	8	6	3	10	3	4	22	2	4	6
Sales	3	1	2	2	1	5	0	2	5	1
Management	19	28	5	11	7	19	29	13	23	17
Consulting	6	1	2	3	15	2	10	2	11	14
Teaching	3	4	4	2	2	2	4	1	2	2
Other	9	5	7	5	9	7	8	6	8	8
Base numbers	2250	64	113	294	454	340	251	304	89	85

Source: National Career Development Project, Purdue University, W. Lafayette, Ind., 1982.

CONSTRUCTION COMPANY	MANUFACTURING COMPANY	CIVIL SERVICE
Inspector	Engineering assistant	Engineer trainee
Assistant engineer	Junior engineer	Assistant engineer
Resident engineer	Design engineer	Associate engineer
Project engineer	Project engineer	Engineer
Project manager	Section engineer	Section engineer
Construction engineer	Chief design engineer	Principal engineer
Chief engineer	Chief engineer	Executive engineer

Within a given organization these designations would be adequate. However, to be complete a title should specify rank, branch, and function—for example, chief mechanical design engineer, junior refinery operation engineer, or resident highway construction engineer.

Can you visualize the work of each of these people? Do you know some of the duties and responsibilities of each? Do you know the general qualifications for each of these positions? If so, you have a good picture of the engineering profession and are in a position to benefit from your engineering training.

CONSULTING

Consulting is an engineering activity that cuts across the spectrum of functions. For example, a mechanical engineer who specializes in design may establish his own office and offer to prepare plans and specifications for any client who chooses to employ him. An industrial engineer with particular skill in automation of production processes may offer her services as consultant to a variety of small companies. Consulting differs from other engineering employment in the professional problems encountered, the economic opportunities presented, and the qualifications required.

Consulting engineers tend to follow one of two different paths. The general consultant is typified by the civil engineer who deals primarily with the public. He may survey a piece of property for an individual, lay out a subdivision for a developer, plan a sewage treatment plant for a small community, determine the structural requirements of an apartment for an architect, or design a bridge for a highway district. The technical problems may be routine, but providing the necessary performance and safety at a minimum of cost to the client requires skill and judgment.

Electronics engineers with unusual skill in semiconductor device processing are examples of specialists who consult in a limited area—usually at a high technical level. They offer their services to companies that do not wish to maintain their own staffs in this particular field. A manufacturing firm producing components for a satellite control system may run into difficulty in maintaining consistent performance of an integrated circuit at high temperatures and may call for a consulting electronics engineer. Another firm may retain consulting electronics engineers for advice on how microprocessor circuits may accomplish a desired purpose.

Working for themselves, the consulting engineers run a greater financial risk than do the employees of a large company; on the other hand, they have a much greater opportunity to gain full benefits from their skill and effort. On the average, successful consulting engineers earn substantially more than their counterparts in industry. However, with success usually comes growth, and they may find that their attention is directed to business and personnel problems rather than the technical problems they sought in the beginning.

Qualifications for Consulting

Although the qualifications required depend upon the particular consulting area, some general conclusions can be drawn with respect to personality and experience.

In addition to the aptitudes indicated by the technical function to be performed, consulting engineers must possess a special facility for meeting the public and selling themselves and their ideas. In most cases, business sense and organizational ability are just as important as technical skill in determining professional success in this area. Outstanding consulting engineers usually combine high ethical standards and a strong sense of personal integrity with exceptional initiative and unfailing drive.

By definition, consultants are accomplished practitioners. Before hanging out their shingles, they must have had the experience necessary to develop skill and judgment. In addition to the bachelor's degree and graduate work commensurate with the requirements of a functional specialty, they need the type of professional training that can be obtained best under the supervision of highly competent engineers. Practical knowledge of business procedures, commercial law, and personnel supervision are also essential.

TEACHING

Engineering teaching is another activity that cannot be restricted to a single function. Individuals with unusual competence in design, research, construction, or management may choose to exert their major effort toward the instruction of engineering students while conducting research or practicing on a consulting basis. Other opportunities exist to teach engineering or related subjects on the high school, junior college, or technical institute level. Although the typical engineering undergraduate is more interested in professional practice, the importance of teaching is so great and the rewards are so satisfying that it merits consideration here.

It is generally accepted that a university has three basic functions: to provide in its libraries and collections a repository for the knowledge of the world, to transmit this knowledge to earnest students by means of its teachers, and to add to this knowledge through the efforts of its scholars and researchers. Engineering professors make their contributions as teachers in the classroom, as advisers in the office, and as investigators in the laboratory.

Instruction

As a freshman student, you see your professors primarily as classroom teachers. Preparing for their classroom appearances, they spend time as scholars, planners, and educational psychologists. All teachers are engaged in a continuous learning process; they must constantly enlarge their horizons, increase their depth of understanding, and keep up with technical advances. In planning the best utilization of their time and that of their students they must review, select, and organize the appropriate information. Their courses should take advantage of the educational possibilities of textbooks, lectures, demonstrations, laboratories, computers, and other aids such as movies and television. They must challenge the best students and encourage the less capable.

Advising

One of the privileges of teaching is the opportunity to work closely with alert, optimistic, and energetic young people who, while building their own foundations, inspire their elders to higher achievement. The powerful influence of a stimulating teacher on a receptive student has been acknowledged by many leaders in the engineering profession. As a sympathetic adviser and friendly counselor, the teacher can share in the hopes and aspirations of his or her students and help in the realization of these dreams by providing academic assistance, personal advice, and vocational guidance.

Research

The engineering school has an obligation to push out the frontier of knowledge in addition to passing on the learning of a previous generation. As part of personal advancement, one creative teacher may be investigating unsolved problems in her field, gathering data on interesting new phenomena, and making findings available to all through technical talks and published papers. Another teacher whose experience and training provide special competence may be serving as a consultant to industrial concerns or governmental agencies. Still another teacher's contribution may be that of writing a textbook that organizes new advances in a logical and teachable fashion.

The combination of research, consulting, or writing with instruction provides deep satisfactions for the teacher and inevitably improves his or her teaching because it brings enthusiasm, originality, and authority into the classroom.

The Future of Teaching

Due to the rapid increase in the number of young people interested in engineering and technology, there continues to be a significant demand for engineering teachers. An annual demand in the 1980s of 1500 to 2500 additional teachers will be felt in engineering education alone. Salaries of college teachers, traditionally lower than those of capable professionals in other activities, have improved somewhat in recent years; base salaries for 9 to 10 months are usually augmented by summer employment, consulting, public service, and patent or writing royalties. Greater public recognition of the importance of education coupled with the facts of supply and demand assure an improved position for teachers, both economically and socially speaking.

Teaching is a complex activity involving the communication of abstract ideas and personal relationships between individuals. The good teacher has an opportunity to influence his students in his office as well as in the classroom and laboratory. Exchanging ideas with young people is one of the most stimulating aspects of teaching.

Of far greater importance, however, is the assurance that teaching will continue to provide rewards and satisfactions not found in any other activity. Working with tomorrow's leaders and contributing to our fund of knowledge, the engineering teacher never needs to question whether he or she is doing something worthwhile. Following self-selected avenues for investigation and providing personal standards of performance, the teacher has an unusual amount of freedom. Living and working in a community of scholars where creativity is honored more than possessions and where opportunities for intellectual and cultural stimulation abound is another positive advantage of an academic career.

Qualifications for Teaching

But teaching is not for everyone. To teach a subject requires a high degree of mastery and an ability to communicate; these are not common. Therefore, a high intellectual capacity is necessary; perhaps ranking in the top one-quarter of one's class is a good indication of this. Effectively communicating abstract principles and complex concepts to young people and satisfactorily replying to their cross-examination call for a special ability in oral expression. Providing one's own direction and setting one's own criteria require initiative, enthusiasm, and perseverance.

The normal route to engineering teaching includes graduate work to the doctoral degree, which means approximately three years of full-time work beyond the bachelor's degree. The doctoral degree is awarded in recognition of broad mastery of the basic engineering and physical sciences, special competence in a particular field at an advanced level, and ability to carry out original research and present the results. There are usually opportunities for part-time teaching while pursuing the advanced degree. The route is long and the way is hard, but for one who has an urge to learn, enjoys mastering new subjects, and finds satisfaction in helping others to enlarge their horizons, teaching offers extraordinary opportunities.

SUMMARY OF ENGINEERING FUNCTIONS

Figure 6-4 (page 155) provides a convenient method for relating the various engineering functions.

Research

Research engineers are seeking new knowledge or a better understanding of known facts. They are seeking new principles, new methods, new processes, new truths. They are working on the frontier of scientific knowledge, and their training must provide a thorough understanding of advanced mathematical and scientific concepts. They must have the ability to reason inductively and abstractly and to express themselves in new mathematical forms. They should be skilled in analysis.

Development

The job of the development engineer is to apply the discoveries and results of research to some useful purpose. Development engineers use the principles, tools, and techniques made available through research. Their efforts usually result in working models of a circuit, process, or machine that have desirable characteristics. The development engineer must be ingenious, creative, and skilled in experimentation and have great initiative and perseverance.

Design

Design engineers take the results of development engineers and carry them to the point where they are economically useful. In designing products to be manufactured and sold at a profit, they select methods of accomplishing desired results, specify materials, and determine shapes to satisfy physical, chemical, electrical, and thermal requirements. Design engineers must have advanced training in the properties and behavior of materials and processes and the ability to adapt recent advances to current practice. They must be skilled in synthesis, proficient in graphical expression, and have a strong background in economics.

Construction

Construction engineers are responsible for surveying sites and preparing locations for structures. They determine the procedures to be followed on the basis of economy and desired quality of the result. They direct the assembly, placing, and joining of materials. They organize the personnel to carry out these operations. Construction engineers must be skilled in the art of their profession. They must be able to direct people effectively and have a firm grasp of costs.

Production

Production engineers are concerned with plant layout and equipment selection, while paying particular attention to human and economic factors. They choose the process of manufacture, the sequences, the tools, and the methods. They integrate the flow of materials and components with the processes. They devise work stations that facilitate human and automated efforts to increase production. They provide inspection and testing facilities and procedures. They eliminate bottlenecks and correct faults in manufacturing procedures. They work closely with designers through the early stages of production and participate in redesign.

Operation

Operation engineers control machines, plants, or organizations that provide services such as power, transportation, communication, data processing, or storage. They are responsible for the selection, installation, and maintenance of equipment. The supervisors of over-all manufacturing operations are sometimes called plant engineers. They are responsible for preventive maintenance programs and the operation of complex equipment for maximum economy.

Sales and Application

Sales engineers analyze the customers' requirements and select and recommend units to satisfy their specifications most economically. They combine salesmanship with technical education and training in application. They analyze the customers' complaints and train the customers' operating personnel. They must be able to deal with persons on all technical levels from maintenance supervisor to research scientist. Application engineers must be skilled in verbal expression and have strong business training.

Management

It is the responsibility of management to determine the main purposes of an enterprise, to anticipate areas of future growth, to select the most promising research projects, and to formulate the policies to be followed. The manager sets up the form of organization and the chain of authority, and selects the executive personnel. Engineers have proved to be valuable in management positions because of their ability in analyzing the factors involved in a problem, collecting the necessary data, and drawing sound conclusions. The primary requirements include a background of engineering fundamentals, skill in human relations, and ability in business. In the final analysis, the future of the company depends on capable management.

REFERENCES

1 National Society of Professional Engineers, "Income Salary Survey, 1981," Washington, D.C., June 1981.

2 Newcomer, Mabel, "The Big Business Executive—The Factors That Made Him: 1900–1950," Columbia University Press, New York, 1955.

3 Atelsek, F. and Gomberg, I., "Recruitment and Retention of Full-Time Engineering Faculty," American Council on Education Panel Report 52, Washington, D.C., October 1981.

4 "Design for Tomorrow, 50th Anniversary Edition," *Heating, Piping/Air Conditioning,* September 1979.

5 "A Wide World of Careers for Chemical Engineers," *Chemical Engineering,* January 2, 1978.

6 National Science Foundation, "The 1972 Scientist and Engineer Population Redefined, Volume 1. Demographic, Educational, and Professional Characteristics," May 1975, pp. 75–313.

7 Frauenfelder, Judith L., Shalliol, Will, LeBold, William K., "A Quarter of a Century of Questions and Answers, The Purdue Experience," 1981 College-Industry Education Conference Proceedings, ASEE, Washington, D.C.

8 National Society of Professional Engineers, "Income and Salary Survey, 1981," June 1981.

9 Annals of Engineering Education, American Society for Engineering Education, "A Comparison of Men and Women Undergraduate and Professional Engineers," December 1981.

ASSIGNMENTS

1 It is said that the branches usually relate to an engineer's interests, whereas the functions tie in more closely with his attitudes and training. In a paragraph or so, discuss this statement and point out specific examples that support or contradict it.

2 List what you feel to be your own personal characteristics and aptitudes, and indicate the function for which you presently feel best qualified.

3 List three courses at your school that would be particularly appropriate as preparation for (a) research, (b) design, (c) construction, (d) management.

4 If consulting engineering were to be placed on the spectrum of Fig. 6-4, where should it go? Explain why.

5 Select an article (such as a transistor, an airplane wing, a desk-top computer, a gas turbine, or a plastic material) and familiarize yourself with all phases of the manufacture and use of the article. (Some library study may be necessary.)

   a Mentally, classify the engineering performed on this article by function.

   b Describe, in a brief report, the work done in connection with the selected article in each of three engineering functions. This can be done by writing a paragraph on each or by listing in outline form.

6 Referring to Fig. 6-5, what are the distinctive personal and educational qualifications required for work in: *(a)* area *A, (b)* area *B, (c)* area *C?*

7 List two similarities and two distinctions between the functions of (a) research and development, (b) development and design, (c) design and production, (d) production and operation, (e) sales and management.

8 Successful management requires that the needs and values of employees be recognized in setting tasks and in giving rewards.

   a In what respects would the needs and values of research engineers differ from those in construction?

   b In what respects would the needs and values of design engineers differ from those in sales?

9 Select an engineering project with which you are familiar, or discuss with an engineer a project in which he or she was involved. Describe in some detail the work done by an engineer performing a specified function on the project.

10 Get the latest engineering salary survey of the Engineering Manpower Commission and compute the percentage of people in supervisory positions and compare this with the data in Figure 6-3. Has the percentage increased or decreased? How would you explain the constancy and/or change?

11 Construct the linear spectrum of engineering function shown in Figure 6-5 as a circle, assigning the percentages to proportional sectors. Does that configuration explain the interrelationships between teaching and research and teaching and consulting any better? What about the relationships between other functions? Can you design any other configuration that could explain the functional relationships any better?

CREATIVITY AND
DECISION MAKING

In a broad sense, the essence of engineering is design, planning in the mind a device or process or system that will effectively solve a problem or meet a need. The preceding chapters have discussed the achievements of engineers in solving problems (classified by branches) and the characteristics of the problems encountered (classified by functions). In any significant problem, the engineer must create possible solutions and decide whether they are satisfactory. A discussion of these mental processes of creation and decision making should provide a better understanding of the nature of engineering.

CREATIVITY

The engineer tackles only new problems; problems that have been solved before are usually turned over to calculators—human or electronic. The new problems are sure to be difficult because all the easy problems have already been solved. In the search for a solution, the engineer is in competition with a rather intractable environment, with predecessors who tried and failed, and with eager rivals who at that very moment are endeavoring to solve the same problem. An engineer's success depends on the ability to

167

come up with a new idea, a new technique, a new process, or a new material; in other words, it depends on his or her creativity.

The emphasis on new, fresh approaches in a competitive atmosphere makes engineering an exciting profession. The difficulty of the problems, the necessity for abstract thinking, and the emphasis on predicting future behavior make engineering a challenging profession. The elation that comes with the successful solution of a difficult problem in an original way makes engineering a satisfying profession, particularly for the creative individual.

Examples Of Creativity

Imagination and innovation, when coupled with motivation and based on knowledge, lead to advances in science and technology.

The Van Allen Radiation Belts   The early U.S. satellites detected the presence of concentrations of charged particles in the space surrounding the earth. With just a few clues to work with, a team at the State University of Iowa headed by J. A. Van Allen conceived the possibility of immense belts of charged particles temporarily trapped in the earth's magnetic field. They then proposed experiments that later confirmed the existence of the zones shown in Figure 7-1.

FIGURE 7-1
A section through the earth's radiation belts. (*McGraw-Hill Encyclopedia of Science and Technology.*)

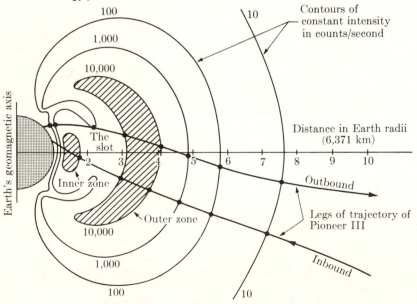

The inner zone contains electrons of energy greater than 600,000 electron volts and protons of greater than 40 million electron volts. These are believed to be the products of the radioactive decay of neutrons moving outward from the atmosphere where they were created by cosmic rays. The outer zone contains electrons of energy greater than 200,000 electron volts and protons of greater than 60 million electron volts. The outer zone is probably due to ionized solar gas that left the sun and became trapped in the earth's magnetic field.

Aid for a Tired Heart   The skill of the engineer in devising pumps, valves, controls, and communication systems has been brought to bear on the problems of diseased or worn-out human parts. Synthetic replacements or prostheses are commonplace in certain situations, and artificial kidneys and heart valves are becoming widely used.

The problems in placing a two-chambered device capable of pumping 7 liters of blood per minute within the chest cavity and maintaining its operation reliability for an indefinite time are enormous. The materials used must have special mechanical, chemical, and electrical properties so that they are not rejected by the defensive mechanisms of the human body. As an essential element in a vital process, a satisfactory prosthetic device must be completely reliable.

The cardiac pacemaker represents a successful solution of one difficult control problem complicated by the possibility of corrosion and by the need for an infallible power supply. Electrical impulses generated in the pacemaker and transmitted to the heart through electrodes implanted in the muscle tissue stimulate a defective heart into rhythmically contracting sixty beats per minute. Pacemaker implants have now become a routine operation permitting thousands of people to continue a normal life.

Gears Shot from Guns   Gears are customarily made by removing material from blanks in a milling operation. The speed at which material is removed is limited by the heat generated in overcoming cutting resistance. In many situations, frictional forces tend to increase rapidly as the speed of cutting increases, and there is an upper limit to the rate of removal of metal by conventional methods.

At a Siberian research institute, however, Russian engineers have developed super-high cutting speeds. In one experiment they fired a gear blank through a gun and against a ring fitted with cutting tools. Surprisingly, they found that above 100 meters per second, the cutting resistance decreased as the speed increased. At a cutting speed of 750 meters per second, the tool wear on difficult-to-machine materials dropped to 1.5 percent of the usual value.

Creative individuals are the key ingredients of a high technology society. Our greatest resource is the group of men and women with the imagination, insight, and ability to use science and technology to solve the technical problems that face our world. Aeronautical engineer Paul MacCready is a famous example of this select group. In 1977, he became known as the "father of human-powered flight" when his *Gossamer Condor* made the first sustained, controlled flight by a heavier-than-air craft powered solely by its pilot's muscles and won a $100,000 prize that had gone unclaimed for 18 years. Following an ingenious set of experiments and employing strong and light space age materials in optimal configurations, he created a 70-pound craft that required only one-third horsepower for propulsion. His second-generation *Gossamer Albatross* was pedaled across the English Channel in 1979, and his sun-powered *Solar Challenger* (shown here) made a 92-minute flight in 1980. (*The duPont Company*)

The Creative Environment

The foregoing examples illustrate the results of creativity: a new concept of interplanetary space, a new device for a hostile environment, a new process taking advantage of an unpredicted behavior. What common elements are found in these examples? What are the ingredients of creativity?

One obvious essential is intelligence. Creativity is a mental process and the abstract reasoning required is closely associated with intellectual ability. But intelligence is not enough; in fact, those persons who score highest on standard intelligence tests are not the most creative. It appears that all persons of above-average intelligence (as a college student, you are in this category) are creative to some degree. A second ingredient of creativity is knowledge—knowledge of the factors and processes involved in a problem, usually resulting from training and experience. A third essential is motivation; the creative person usually has a strong urge or desire to find a solution to a challenging or perplexing problem. Contributing to motivation is the stimulation provided by the group of intelligent, trained, and motivated persons making up a design team or found in a research laboratory.

The Creative Process

Early in 1945, Charles H. Townes of the Bell Telephone Laboratories wrote in an internal memo:

> Many difficulties in manufacture of conventional circuit elements can be obviated by using molecules, resonant elements provided by nature in great variety.... Molecular gas may be heated or otherwise excited to emit discrete bands of radiation.... Although this application probably warrants investigation, it does not appear immediately promising.[2]

He went on to explain that on the basis of thermodynamic reasoning the expected radiation intensity would be very small. In spite of this unfavorable aspect, he was attracted to the very high frequency range between microwaves and the infrared, "because this no-man's land could be predicted to be rich in spectra and because it represented a challenge."

Six years later he could look back on his own work at Columbia University and that of others and conclude that some progress had been made but that good sources of very short waves seemed far in the future.

> In the spring of 1951, I was in Washington to attend a meeting, and I found myself sitting on a bench in Franklin Park early one morning, admiring the azaleas, then at the height of their bloom, but also wondering whether there was a real key to the production of very short electromagnetic waves. Suddenly I realized what was needed.
>
> Clearly we must rely on molecular resonators, since the appropriate elements

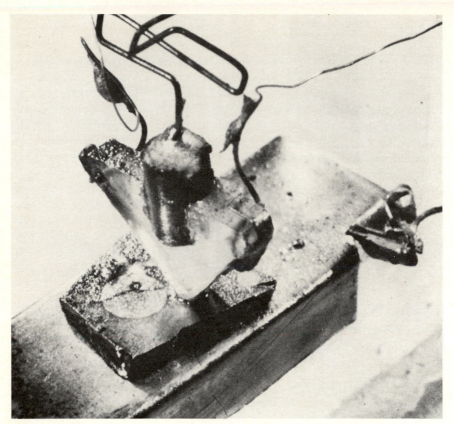

After World War II, emphasis on semiconductor research accelerated at Bell Telephone Laboratories. Experiments led to new theoretical concepts, and William Shockley proposed an idea for a semiconductor amplifier that would critically test the theory. The actual device had far less amplification than predicted, and John Bardeen suggested a revision of the theory. In the next experiments, Bardeen and Walter Brattain discovered a new phenomenon and created an entirely new device—the original point contact transistor shown above. Their epoch-marking experiment is described in Brattain's laboratory notebook. The next year Shockley invented the junction transistor, the forerunner of most modern transistors. (*Bell Telephone Laboratories*)

are too small for man to construct. The thermodynamic argument was faulty because it implied equilibrium, a state easily obviated by recently demonstrated techniques. The feedback necessary for oscillation could be provided by a cavity containing the radiation. In a few minutes I had calculated, on the usual back of an envelope, the critical condition for oscillation in terms of the number of molecules which must be supplied and the maximum allowable losses in the cavity. It was very exciting to find in principle a way of producing a very short-wave oscillator even though it did not look easy or promising.

Two specific things had previously been lacking: a mechanism for obtaining net amplification and strong motivation for attempting what seemed to be a very difficult experiment. . . . It should be remembered that, although this new type of amplifier was an interesting idea, few regarded it as a "hot" topic for either physics or engineering . . . . [To indicate] how uncertain our steps toward progress can be, I might mention a visit from two of my most distinguished colleagues who urged me to stop this silly work on a molecular oscillator because it was wasting government money. Their objections were a help, because the natural result was that I pushed on the experiment harder than ever, and in late 1953, the maser oscillated.

And microwave amplification by stimulated emission of radiation had been achieved.

As Townes points out, scientific knowledge is fruitfully accumulative; one idea is added to another, making a third still easier to understand. The overall structure of knowledge becomes impressive, even though the individual steps are small. The invention of the maser is a good example of the process whereby one creative individual takes an important, and not so small, step along the path of progress.

The Creative Individual

Innovation consists of organizing elements into new relationships; the result of creativity is an original design, an unpredicted theory, a unique pattern, or a novel configuration. The creation of a new composition is an activity common to the artist, the musician, the poet, the scientist, and the engineer. Knowing what innovation is and having looked at some examples of creativity, what can we do to become more creative?

Creativity may be our greatest natural resource, and it may be our least developed resource. With machines replacing muscles and with computers performing routine information processing, innovation takes on increased importance as a truly human function. A great deal of time and effort are being devoted to determining how this resource can be developed—how creativity can be enhanced.

One approach is to study the personality traits of individuals who are identified as being creative. From the many studies of this kind we do get a picture of typical innovators (realizing, of course, that truly creative men or women are never "typical"):

They are sensitive to problems; they are aware of things as they really are; they are unimpressed by status symbols; they are independent and say what they think.

They accept uncertainty, ambiguity, conflict, and change; they can stand frustration and failure; they are open to suggestion; they are willing to change their minds, and then change back again.

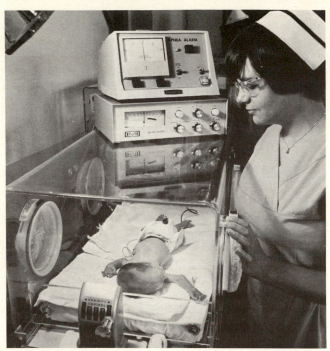

Innovation consists of organizing existing elements into new relationships. Apnea, a sudden cessation of breathing, occurs in approximately 30 percent of premature infants weighing less than 5¹/₂ pounds and may cause brain damage or death. This monitoring device, created by a team of pediatricians and engineers, incorporates a warning alarm system based on fluctuations in current between electrodes attached to the baby's chest. If the infant stops breathing for 20 seconds, the alarm sounds and the attending nurse can start a simple treatment to restore breathing. (*Stanford Medical Center*)

They are intelligent, curious, and skillful in dealing with abstractions; they are able to transform ideas into new meanings; they like to play—to toy with ideas and to clown around with concepts.

They are able to integrate, to put things together, to synthesize; they are able to perceive patterns and to extrapolate from experience to new situations.

They are self-directed and highly motivated; they are zealous and aggressive; they have determination, perseverance, and energy.

If you compare your own traits to those listed, you will probably find that you possess all of them to some degree. What does this mean in your own development?

Developing Your Creativity

There is convincing evidence that innovation is a learned behavior and that creative skill can be developed and improved. Rather than an inherent aptitude, creativity probably reflects an attitude, and attitudes can be changed. Some colleges offer courses in creative thinking; however, your own program of originality development may be more effective. Here are some general suggestions:

Learn more about creativity. Take an interest in this intriguing subject, read about it (the references listed will give you a start), and talk to others about it.

Expose yourself to ideas outside your own special interest; rub minds with others. Play with ideas, twist them about, turn them inside out (and shake them a little).

Take time for leisurely thinking, for contemplation, for daydreaming. Suspend action, and defer judgment; give your ideas a chance to incubate. Enjoy the elation that comes with new insights.

Put yourself in a problem-solving mood, ask questions, strip away the unimportant details, cut through the conventional barriers, and try to define the central issue.

Force yourself to take a new look at old questions. Take advantage of the horizon-expanding techniques of Adams and Gregory (see the references).

Above all, however, appreciate the fact that you have far more creative potential than you have ever used. Developing that potential is worth careful attention because the society of which you are a part values creativity highly and rewards it generously. Even more important, innovation is a high-level human function and a source of great joy and satisfaction.

DECISION MAKING

At 5:14 on the afternoon of November 9, 1965, life was routine for the 30 million people in the 80,000-square-mile area constituting the northeastern United States. In New York City, for example, lights were on in homes and offices, radio and television sets provided immediate contact with the outside world, unseen heating and refrigerating systems hummed unobtrusively, babies were being born and lives saved in closely monitored surgery rooms, a flood of products of all kinds moved smoothly along continuous production lines, elevators lowered their human cargo swiftly and surely within skyscrapers, and millions of people sped safely toward their destinations by controlled surface, underground, and air transportation.

At 5:15 P.M., a small current increase tripped a backup protective relay on one of five transmission lines running north to Toronto from the huge Beck Station of Ontario Hydro on the Niagara River. The relay opened circuit breakers, which removed the heavily loaded transmission line from the system and shifted its burden to the other four lines, overloading them. Protective relays on these lines responded and within a few seconds the total power output of the spinning generators surged into the interconnected transmission lines leading south to New York State. In the next few seconds there was a cascading series of power inrushes, load drops, frequency changes, phase shifts, and current surges that triggered protective devices over the entire area. The finely balanced Canada-United States Eastern Interconnection system broke down into a series of isolated segments, most of which were unable to marshal spinning reserves rapidly enough to meet load demands.[1]

Plant after plant was shut down to prevent destruction of generating equipment, and in New York, hundreds of miles away, the lights dimmed and flickered out. Inhabitants of that great metropolis were trapped in stalled elevators, powerless subway trains, massive traffic jams, or the dark confines of their own apartments. In just a few minutes, the beneficiaries of the most advanced technology the world has ever known were reduced to groping about in the dark with candles.

How could this inconceivable technological paralysis occur? What incorrect decisions were made? Who was responsible? Was this catastrophe due to lack of knowledge or to the use of improper criteria?

Control of a complex power system is certainly an engineering responsibility. The engineer not only makes technological advances available to the world, but must also see that such advantages do, indeed, enhance its welfare. By virtue of his or her knowledge, skill, and unique role in society, the engineer is best qualified to make the technical decisions that make technological progress possible. In the role of decision maker, the engineer must be concerned with the results of that progress. Let us examine some of our major problems to see what role the engineer must play in the future.

Technical Decisions and Human Welfare

What can be learned from the 1965 New York blackout? It now appears that the technology for power-system control was available but that errors in engineering judgment were made in not putting sufficient emphasis on reliability. Systems cannot grow by merely connecting existing systems; increased size and complexity require more sensitive controls and more elaborate reliability provisions. In arriving at design decisions on a system that reaches into every corner of our technological society, the engineer must give consideration to reliability and aesthetics as well as cost and safety.

In defining engineering, we emphasize the optimum use of natural

---

[1]"The Great Power Blackout and Engineering Responsibility," *American Engineer,* April 1966.

resources for the benefit of humankind. The process of optimization implies a careful selection among alternatives and engineering is essentially a series of decisions. Technical decisions involving quantitative factors are relatively simple where direct numerical comparisons can be made. Where human beings are involved, as they are in all important engineering problems, qualitative factors become significant and the engineer's decisions are far more difficult.

Let us now look at one broad area where engineering decisions will be required, and identify the factors that must be considered. Selected for this closer look are the problems of control of the human environment. Our ultimate objective is to describe the role to be played by the engineer of the future and to identify the characteristics needed to be successful in that role.

The Human Environment

A major aspect of our early space exploration program was an attempt to determine if life exists on other planets. The first living things appeared on earth nearly 500 million years ago, but it was only about one million years ago that the reasoning creatures which we call *Homo sapiens* began to walk on earth. And it is only within the past decade that man has had the knowledge and the tools to undertake a thorough study of neighboring planets.

We know quite a bit about human history from the evidence left in caves and around campfires. The creatures that preceded humans were at the mercy of their environment: They hid from the sun, migrated from the cold, suffered from the wind, avoided the rain, and fled from fire. Humans, however, conquered their environment, learned to control it, and turned it to their advantage. As masters of their environment, they seek the sun at the beach, play in the snow in the mountains, use the gentle winds on a lake or the jet streams high in the sky, take advantage of the rain and harness its runoff, and kindle fire for comfort. But humans have gone far beyond nature. They have created and used "fires" that rival the sun in their temperatures and cold that approaches the infinite cold of absolute zero. They have created winds of far greater velocity than nature generates in the wildest storm and used them in routine cross-country travel. They have provided life-giving water where adequate rain has never fallen and thereby turned deserts into farms.

One clear distinction between *people* and all other living creatures is this ability to understand, control, and modify the environment. As a result we have surrounded ourselves with a complex assemblage of products, structures, tools, institutions, and customs that stand between us and nature. We have created our own environment. In doing so we have become the most influential factor in our own evolution. While achieving extraordinary freedom from nature, we have assumed the burden of responsibility for our own welfare and that of generations to come. In applying science, converting resources, and creating things, the engineer plays a key role in environment modification.

Cities

The earliest people were nomads, who hunted animals and gathered food where they could find it. About 10,000 years ago, the pattern of human living suddenly changed. Humans became food-producers, domesticated animals, cultivated crops, and stayed in one place. With an assured supply of food, people had time to develop other energy sources and the accompanying technology; they began to modify the environment and to significantly alter the natural world.

The clustering of human beings into cities appeared spontaneously in widely separated regions and seems to have been in response to basic desires. In Aztec, Mayan, Indian, Chinese, and Egyptian civilizations, cities were the centers of development and change. The history of ancient Europe is the story of Rome and the Greek city-states. The dynamic life of the Middle Ages centered in such active cities as Milan, Paris, London, and Hamburg. However, the modern city is a product of the industrial revolution, which brought about a concentration of power, labor, manufacturing, commerce, and finance.

By 1980, three-fourths of the American people were squeezed together on less than 1 percent of our land. Today over one-half of the total population is living in the thirty largest cities. The trend is toward huge metropolitan areas—the northeast corridor extending from Boston to Washington, the region south of the Great Lakes, and California between San Francisco and Los Angeles. These supercities will pose environmental problems of unparalleled magnitude and complexity and provide correspondingly great opportunities for imaginative planning and creative execution.

Why do we choose to live so close together? With our affluence and mobility we can live almost anywhere we like. We no longer need the protection of the fortress-city; in fact, the inner city is becoming an increasingly hazardous place in which to live. With advances in communication, the resident of a rural community can avoid the smog, the ugliness, the traffic jams, and the riots, and still not feel isolated. What pulls us to the city?

Perhaps it is because the city provides the most intense human interaction. At the crossroads we find an exchange of people as travelers come and go, an exchange of commerce as raw materials and finished products move in and out, and an exchange of ideas among dynamic people who want to be where the action is. The city is a center of action, change and, therefore, opportunity; job opportunities in urban areas, for example, are increasing even faster than urban populations.

Young people and members of minority groups see the city as a place of opportunity and are attracted to it. At the same time, talented and affluent individuals see it as a place to be tolerated during working hours. Some of our major urban problems arise because of the vast numbers who wish to spend the day in the stimulating atmosphere of the inner city and then escape to more attractive suburban surroundings. Compounding these problems are the large numbers who work in factories on the fringes of the city and return each night to downtown tenements.

The City as a System

What can be done to make the city an attractive place in which to live—a place to be enjoyed instead of a place from which to escape? How can the virtues of a city be enhanced and the problems be mitigated? Much of the necessary work lies in the province of the engineer. At present, the great emphasis is on improving transportation, but that may just be a symptom of a greater inadequacy. Future emphasis must be on living and not just on moving about. Improvement of public health and safety, substitution of beauty for ugliness, and elimination of pollution of all kinds are certain to be of great public and direct engineering concern for the next decade.

An appreciation of urban problems can be gained by applying an approach called the systems concept. In engineering, a system is a collection of matter within specified boundaries; the behavior of the system is described in terms of input and output variables. Ideally, input quantities are governed by events occurring outside the system; the output quantities are determined by events occurring inside the system in response to inputs. A pumping system operated on an isolated farm is shown in Figure 7-2. Air, fuel, and water are readily available as inputs; combustion products are exhausted to the atmosphere, and the desired output (water under pressure) is supplied to the house and orchard.

FIGURE 7-2
An open-loop pumping system.

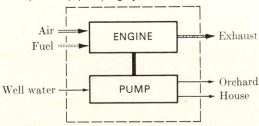

For an isolated farm, this is an adequate model because the outputs have negligible effect on the inputs. However, if a hundred farmers live in a single valley or if a million people are gathered together in a city, a different type of model is required. As shown in Figure 7-3, the atmosphere provides a link between input and output air, and the ground "closes the loop" for the water quantities. (The orchard products in passing through the marketplace generate funds for the fuel.) The behavior of such a *closed-loop system* can be quite complex because a change in output causes a change in input that may cause a further change in output. Such systems have certain advantages and are widely used in engineering; however, they are likely to become unstable unless adequate controls are present.

It is impractical for each urban dweller to provide his or her own water, sewage disposal, or electricity. Instead each dweller's house includes parts of a water system, a sewage system, and a power system. In addition, there is a postal system, a telephone system, a refuse collection system, a fire and police protection system, and a transportation system. Also, there usually is a park system, a recreation system, a book circulation system (library), and an art distribution system. Of particular importance to engineers are those complex, large-scale systems in which proper performance, safety, and efficiency can be obtained only through correct technical decisions.

Trade-offs in Decision Making

It is clear that we can to a large extent design our own future, but what is best for the members of a bright new community? Looking back over the

FIGURE 7-3
A closed-loop pumping system.

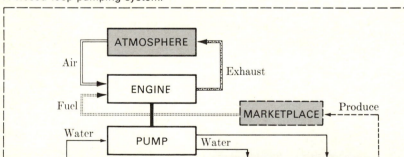

problems described in this chapter, a long series of basic questions that require trade-offs in the decision-making process come to mind:

In designing electric power systems, should emphasis be placed on low cost or high reliability?

To what extent should low power cost or high system reliability be sacrificed to improve appearance as in placing distribution lines underground?

Should population growth be accommodated by creating new cities on rural sites or by expanding and improving existing cities?

Should cities be subdivided into smaller communities by freeways and greenbelts or should the emphasis be on integration into metropolitan areas?

Should zoning ordinances and building codes be modified to permit architectural and artistic control of all new structures?

Should every automobile include in its purchase price a "funeral fee" to take care of its eventual disposal?

Should the major portion of public funds be devoted to eliminating river pollution or to developing new sources of water? Should the use of potent new chemicals such as DDT, which can virtually eliminate the malaria-carrying *Anopheles* mosquito, be halted until the long-range effects on beneficial insects are established?

Should long-lasting chemical sprays be prohibited in agricultural uses where significant amounts of spray material may enter the atmosphere or appear in streams draining the area?

Should disposal of sewage by water flushing be prohibited in new housing units?

Should automobile engine efficiency and performance be sacrificed in order to eliminate emission of pollutants?

Should the largest portion of the research effort of a private petroleum company be invested in reducing the cost of gasoline, in improving the quality of gasoline, in decreasing the pollution emission of automobile engines, or in developing a pollution-free fuel cell?

Should public funds be devoted to the development of new pollution-free automobile propulsion systems?

THE FUTURE ROLE OF THE ENGINEER

Two distinguishing characteristics of the modern era are the extent to which technology is effective in producing change and the speed with which the world is changing. As translators of science, engineers should play a decisive role in determining the future course of human existence. If they are to be given the responsibility for which their technological competence uniquely qualifies them and if they are to discharge that responsibility well, engineers must be prepared to make technical decisions of far broader scope than those solved successfully by their predecessors.

The Influence of Science and Technology

Every aspect of modern life is influenced by the application of scientific discovery to basic human problems. In one sense the dominant position occupied by new creations—automobiles, computers, assembly lines, bridges, and skyscrapers—poses a threat to the individual himself. On the other hand, society's creations place within its grasp for the first time the possibility of a life free from poverty, disease, ignorance, and drudgery.

In addition, we can free ourselves of many of the constraints that have heretofore been accepted as inherent in human existence. Already we can control our immediate physical environment and are nearly free from the restrictive effects of nature's heat, cold, rain, or wind. We remove the threat of flood and increase productivity by rerouting streams, creating lakes, clearing forests, or filling in bays. We overcome the restraints of distance and time by crossing continents routinely and performing a year's mental calculations in a few seconds of computer time. No longer restricted to the thin "skin of the apple" on which we reside, we are planning voyages to the bottom of the sea and into interplanetary space.

As we have seen, however, technological advances may have unpleasant side effects. Much of our mobility depends on the cheap and efficient gasoline engine that pollutes our atmosphere. The toxic sprays that increase agricultural productivity by killing harmful insects may be washed into streams where they kill beneficial creatures as well. The new industrial robots that benefit the majority by increasing production quantity and quality may bring misery to a few by eliminating their jobs. The efficiently designed transmission line or highway may spoil the appearance of a hillside or destroy the magnificence of a centuries-old stand of trees.

The power of technology in controlling the human environment may become even greater in the near future. Formerly an engineer was concerned with only a single aspect of the environment. The automobile, the TV set, the contraceptive pill, or the teaching machine could be a strong influence on human behavior, but it was only one of many influences acting somewhat randomly. Now, however, huge industrial firms with vast resources of capital, management, research, and technological know-how are entering the social area. For example, a major electrical manufacturer intends to design and construct over a twenty-year period an entire city for a future population of 100,000 people. The company expects to develop new concepts of urban living and new building materials, components, and procedures. It intends to design new transportation, traffic control, power distribution, waste disposal, police, hospital, information, and education systems. It sees this billion-dollar project as a challenging outlet for its sophisticated technological and managerial talent; also it envisions the creation of vast new markets for its manufacturing capacity. The citizens of such a city, for good or bad, will be subjected to an environment distinguished by its consistency with a single planning theme.

The Engineer of the Future

Our society is future-oriented. Our culture and environment are changing rapidly in fundamental ways, and we must plan ahead. We need people who can anticipate technological developments, predict their social

An increasing number of women will be involved in future engineering projects. Here, a woman mission specialist for future space shuttle flights checks the fit of a flight suit in the cockpit of the space shuttle orbiter flight-deck simulator. The simulator provides training for the flight crew in monitoring command and control capabilities required for the management of all normal and emergency flight operations.

consequences, and make the decisions that will result in the maximum long-range benefit for the majority with the minimum short-range disadvantage for the minority. Where will these people be found?

The intimate relation between scientific and technological advance and social and environmental change indicates the possibility of a grand new role for the engineer. The need is for professional persons educated in society as well as science, concerned with human beings, and capable of applying science in the creation and operation of complex socio-technological systems. What will be the characteristics of these engineers of the future?

They will be willing to assume the responsibility for decision making. This is a difficult role and in the past most engineers have avoided this kind of responsibility. They have limited themselves to supplying the technical data on which others have based decisions. But where technological factors are crucial, as they are certain to be in the foreseeable future, engineers must be willing to take a stand. For them, failure to do so would constitute unprofessional conduct.

They will be willing to participate in policy making and standard setting. We must ensure that the potential benefits of science and technology are fully realized for the social and economic welfare of all our people and those of less well endowed or less developed nations. We must draft and implement laws to ensure the highest use of natural and environmental resources. We must set standards of safety, reliability, purity, and beauty so that we can be proud of the altered world we pass on to future generations. In addition to participating in constructive actions, the engineer must speak out against misuse of our resources, even to the point of dissenting with the management of his company or the officers of his professional society.

They will have the courage to act on inadequate information in pursuing goals that are not well defined. Traditionally, the engineer has been noted for a cautious approach and reliance on past experience or careful testing. The important engineering problems, however, involve the unknown future and decisions must be based on the probable effect of uncertain factors in unfamiliar situations. After exploiting scientific knowledge and technological skill, the engineer must be willing to push ahead, take risks, and make bold decisions.

The successful engineers of the future will be prepared for decision making on complex problems in broad areas. They will be skilled in the application of sophisticated tools and innovative in the development of new techniques. They will be able to take full advantage of the computer in gaining new insights as well as in computation and information processing. They will understand the decision-making process under conditions of uncertainty. They will appreciate the role played by economists and sociologists and be able to coordinate their work with that of social scientists. They will have the vision to conceive vast projects, the skill to analyze them as integrated human-machine-environment systems, and the ability to predict their technical performance and their human impact.

REFERENCES

Creativity

1 Adams, J. L., *Conceptual Blockbusting,* San Francisco Book Co., San Francisco, 1976.
2 Arieti, Silvano, *Creativity, The Magic Synthesis,* Basic Books, New York, 1976.
3 Gregory, S. A., *Creativity and Innovation in Engineering,* Transatlantic, 1973.
4 Gruber, Howard E., *Darwin on Man: Psychological Study of Scientific Creativity,* University of Chicago Press, Chicago, 1981.
5 Hanson, Norwood, *Patterns of Discovery: An Enquiry Into the Conceptual Foundations of Science,* Cambridge University Press, New York, 1958–1965.
6 Holton, Gerald, *Thematic Origins of Scientific Thought: Kepler to Einstein,* Harvard University Press, Boston, 1973.
7 Kneller, G. E. *The Art and Science of Creativity,* Holt, New York, 1965.
8 Nickles, Thomas (ed.), *Scientific Discovery, Logic, and Rationality,* Kluwer, Boston, 1980.
9 Nystom, Harry, *Creativity and Innovation,* Wiley Interscience, New York, 1979.
10 Taylor, C. W., and Barron, F. (eds.), *Scientific Creativity: Its Recognition and Development,* Wiley, New York, 1963.
11 Toynbee, A., *Has America Neglected Her Creative Minority?,* Utah Alumnus, February 1962, p. 10.

Decision Making

12 Bell, David E., *Conflicting Objectives in Decisions,* Wiley Interscience, New York, 1977.
13 Brinkers, H. S. (ed.), *Decision-Making: Creativity, Judgment, and Systems,* Ohio State University Press, Columbus, 1972.
14 Cetron, Marvin J., *Quantitative Decision Aiding Techniques for Research and Development Management,* Gordon, New York, 1972.
15 Cone, Paul, *Executive Decision Making Through Simulation,* 2d ed., Merrill, New York, 1979.
16 Dickerson, S. L., and Robertshaw, J. E., *Planning Design: The Systems Approach,* Lexington Books, 1975.
17 Dillon, R. J., *Reality & Value Judgment in Policymaking: A Study of Expert Judgments About Alternative Energy Technologies,* Arno, New York, 1978.
18 Goehle, D. G., "Decision Making in Multinational Corporations," University of Miami Research Press, Miami, 1980.
19 Kickert, Walter J., *Fuzzy Theories in Decision-Making,* Martinus Nijhoff, Boston, 1979.

ASSIGNMENTS

1 Write down as many different uses for a common brick as you can in two minutes.

2 You need a machine that will sort 100,000 olives per minute into ten different sizes. (a) Sketch four basically different methods for doing this. (b) Write a one-page description of your thought processes as you worked on part a.

3 A unique water fountain is needed for the entrance to the new research laboratory of a company proud of its creativity. (a) Analyze what fountains usually do. (b) Sketch something entirely different. (c) Write a one-page description of your thought processes as you worked on part b.

4 List five possible nontransportation uses of a bicycle.

5 Devise a new card game to be played by three people using only the twelve face cards in an ordinary deck.

6 As the last thing before going to bed, identify a kitchen problem common to most households and write it down. (a) The next day, propose a solution. (b) Describe your thought processes between the identification and the solution.

7 As the last thing before going to bed, identify a driving problem common to most motorists and write it down. (a) The next day, propose a solution. (b) Describe your thought processes between the identification and the solution.

8 As an exercise in negative creativity, assume it is the year 1890, and give five good reasons why travel by personal automobile is impractical.

9 Discuss whether a computer can be creative.

10 Discuss the effects of formal engineering education on creativity.

11 Automobile accidents are common; airplane accidents are rare. Discuss the similarities and the differences, and the role played by the public in deciding on the characteristics and the operation of these two modes of transportation.

12 In December 1966, Dr. Joseph Shea, Apollo spacecraft program manager, stated that over 20,000 failure reports had been filed on the command and service modules. He said that 60 to 70 percent of the failures were important, but pointed out that they were attempting to run the program in a balanced way. "Don't try to make everything too perfect or you'll never get the job done," he warned. Analyze this as an engineering policy in space exploration.

13 The 4.5-mile rapid transit line in Toronto cost $67 million. Over a ten-year period, over $10 billion was invested in property along the line and the tax income increase directly attributed to the transit line is now estimated at $60 million per year. Assuming that passenger fare income is just equal to operating expenses and stating any other necessary assumptions, analyze this as an investment.

14 It is estimated that 80 percent of high-speed jet travel is for business purposes—that is, to get people from one place to another for face-to-face exchange of information having commercial value. Is some better, less expensive means of providing this type of communication available? Devise and describe, in some detail, an alternative information-exchange system.

15 When the bids were opened on a 21-block section of subway (including two stations) in downtown Oakland, the low bid was $13.5 million over the $48 million

estimate. A quick decision had to be made by the BART management. What are the alternatives? What data would you need in making a decision?

16 A strong conservationist recently said: "Engineers justify dams and highways in wilderness areas by elaborately calculated cost-benefit ratios. What right have they to decide what weight is to be given our social life, environment, aesthetics, recreation, scenic views, and other intangibles?" Discuss this question in about 350 words.

17 Describe three or four specific examples that show how science and technology have become powerful instruments of national policy.

18 An industrial engineer is asked to establish a branch plant in a ghetto. In the plant, automobile panel instruments are to be connected to wires and the wires wrapped into cables for quick installation at the main assembly plant. What social and business advantages do you see in such a location? What problems do you anticipate and what provisions would you make to solve them?

19 The data on an individual's Income Tax Return Form 1040 is recorded on 0.2 inch of magnetic tape by the Internal Revenue Service. At least four other federal agencies may have data on you in their computers. Some day, all the data may be combined in a single computer. (a) Identify the four other federal agencies. (b) What basic safeguards should be enacted into law to protect the rights of privacy of individual citizens? (c) In what other ways will the computer affect us politically?

20 "The questions are always the same; only the answers change." Discuss the meaning of this statement from the viewpoint of an engineer in a decision-making position.

Engineers and technicians
Technologists and scientists
Women and minorities

8

THE
TECHNICAL TEAM

The mechanical development engineer put down the telephone and said to her lab assistant, "The chemist has come up with a new glycol-based ink that will not smudge or dry up the tip of our new high-speed recorder. I want you to set up a reliability test that includes start-stop cycles, speed changes, and input signals of various waveforms."

The nuclear engineer looked over the shoulder of the reactor operator. The log showed a curious increase in radiation level whenever the power exceeded a certain level. "This doesn't look serious," he said, "but I'm going to be talking to the physicist this afternoon and I'll ask her if there is a secondary effect that we're overlooking."

The structural design engineer put down the journal and thought about the new mathematical approach to the analysis of three-dimensional structures he had just read about. When he discussed it with the programmer, she seemed confident that she could computerize the method for use in the analysis of the roof for the enclosed stadium they were designing.

The aerospace engineer studied the sketch prepared by her designer. The camera-operating mechanism looked simple and effective, but the engineer was concerned about the amount of precision required in focusing. Before going further, she decided to check with the physiologist to see how much was known about the effect of weightlessness on an astronaut's ability to make a fine adjustment.

As is indicated in these illustrations, the engineer is usually a member of a versatile team of specialists possessing complementary skills and knowledge. The chemist, the physicist, the mathematician, and the biologist are being employed in industry and government in increasing numbers. Working with engineers, these scientists bring to bear on practical problems the stores of knowledge in their particular fields. Assisting the engineer in the field, on the drafting board, in the laboratory, and on the production line, technicians and technologists make their unique contributions to America's industrial progress.

Frequently, a high school graduate whose real interest lies in the work of a scientist or technologist enters, instead, the field of engineering because the name is more familiar. This is particularly true of the young man or woman whose aptitudes and interest promise success in a career as a technologist but who is unaware of the opportunities in the technical fields closely related to engineering. To explain the relationship between the engineer and the other members of the technical team, to point out the differences in the required training and qualifications, and to describe the employment opportunities available is the aim of this chapter.

THE TECHNICAL TEAM

Engineering includes ideas and action. In preparing for effective service in a particular area of engineering, women and men must concentrate on certain types of training and forgo others. They develop mental skills but seldom have the opportunity to develop manual skills. In concentrating on the application of science they can obtain only a limited knowledge of science itself. As technology moves rapidly ahead, the engineering aspects of any project become more and more demanding. As a result, the most effective organization of technical ability is a team of specialists with each specialist making his or her particular contribution.

Engineering teams vary from small—a few people (for example, an engineer, a technician, and a machinist)—to large—thousands of people in a large public utility or manufacturing company. There are five kinds of technical team members:

1 *Craftworker*—a user of hand and power tools who builds, operates, maintains, or services machines and products useful to the engineering team and other people.

2 *Technician*—an assistant to an engineer or scientist who carries out details of the technical work.

3 *Engineering Technologist*—a practical person who applies engineering and/or scientific principles to the solution of complex problems in a specific area of technology.

4 *Engineer*—a cost-conscious, practical innovator, designer, and problem solver who may be the team leader.

5 *Scientist*—a seeker of basic new knowledge of nature and its laws through research.

Each one is an important member of the team with some persons having the ability to play more than one role on the team. Also, it is important to remember when examining the engineering functions of the various team members, that these functions are all highly interrelated; sometimes it's hard

These pictures show a technical team working on part of a $300 million tracking and data relay satellite system (TDRSS). The system is composed of two spacecraft—one over the Atlantic and one over the Pacific, one dedicated Advanced Wester satellite and one on-orbit spare. The satellites are in geosynchronous orbit and communicate with a single ground station at White Sands, New Mexico. The system will track and relay data between earth and manned and unmanned space vehicles in near-earth orbit. This will eliminate the need for worldwide ground stations, which are costly and sometimes located in politically sensitive areas. The TDRSS incorporates the highest levels of technology available in such areas as communications, guidance and control, and computers (hardware and software). Each $2\frac{1}{2}$-ton satellite will be carried into a 150-mile high low-earth parking orbit by the space shuttle and then raised to 22,300 miles by an inertial upper stage. When deployed it will be as large as a house and its solar panels will generate over 1800 watts of power *(TRW)*

The chief technologist has identified a problem. He knows the characteristics and capabilities of his workers and their machines, but the data in his notebook indicate that the new product will not meet specifications when it gets into production. Drawing on his basic understanding of principles, the engineer suggests a change in fabricating sequence, a substitute for a key ingredient, or a shift in processing temperature. Together they solve the problem. (*Eli Lilly*)

to tell where one role ends and another begins. Decisions or output from one area will greatly influence decisions made in a different area of the engineering process. Also, the five kinds of technical people generally show differences in their personal traits, their work, and their preparation. The craftworker tends to have little formal education but a large amount of on-the-job training as an apprentice-journeyman, whereas the engineer and scientist tend to have a more formal and theoretical education including possibly the bachelor's, master's, and doctor's degrees. The on-the-job training of this latter group is often limited to a short preprofessional experience or a brief orientation to the job and organization.

In attempting to meet its needs, society focuses[1] major attention on different human needs, and at different stages of development new teams of the various types of technical people are formed. The initial stage ordinarily is either a research project seeking basic new knowledge headed by a scientist, or it may be a technical feasibility study headed by an engineer. The second stage often involves engineering development and design led by an engineer assisted by technologists, technicians, and craftworkers. The third

[1] Many of the ideas in this section were developed for engineering guidance by A. Pemberton Johnson and his colleagues as part of a special guidance task force [1].

stage—usually production, construction, and/or operation—is becoming the realm of the technologist who is frequently assisted by technicians and craftworkers, and sometimes accompanied by an engineer who acts as adviser or consultant. The fourth stage of day-to-day operation and maintenance is likely to be conducted by technicians and craftworkers with a technologist or engineer as supervisor.

Whether a person becomes a scientist, an engineer, a technologist, a technician, or a craftworker, and whether she or he remains satisfied in that role, is primarily a matter of temperament and personal interests. The person who searches for general scientific principles of wide significance is more likely to be happy as a scientist; the practical, mathematically-minded, scientifically-informed, active problem-solver alert to costs will probably be happiest as an engineer. The technical person who prefers greater involvement with scientific devices and gadgets than with theoretical concepts will enjoy the work of a technologist. The individual who is less theoretically minded and prefers an intense yet practical involvement in a supportive technical role often becomes an engineering (or physical science) technician. The person primarily interested in using hand or power tools for technical purposes would most likely enjoy employment as a craftworker—a machinist, an electrician, or a mechanic.

Figure 8-1 provides a graphical conception of the relationship between the practical-theoretical orientation and type of education and training of the various engineering team members.

THE CRAFTWORKER

The craftworker typically prefers to work with power or hand tools during the installation, maintenance, or trouble-shooting of physical objects. For example, the industrial electrician uses wires to connect fuses, switches, starters, motors, lamps, and complicated electrical equipment; the automobile mechanic installs, checks, and repairs engines, brakes, clutches, gears, or automatic transmissions. The machinist uses a variety of machine tools to shape metals to someone else's design. The laboratory glassblower is one example of a highly skilled craftworker.

The craftworker who has developed unusual manual skills takes a deep pride in how well he or she can make, install, or repair things. Craftworkers who operate and maintain cranes, bulldozers, and other construction machinery are sometimes called *operating engineers*. Craftworkers usually are employees who use equipment or tools rather than employees who are planners, innovators, or developers of hypotheses. Their full-time education may include postsecondary schooling such as training in a technical institute; it usually involves a four-year formal apprenticeship or informal on-the-job training.

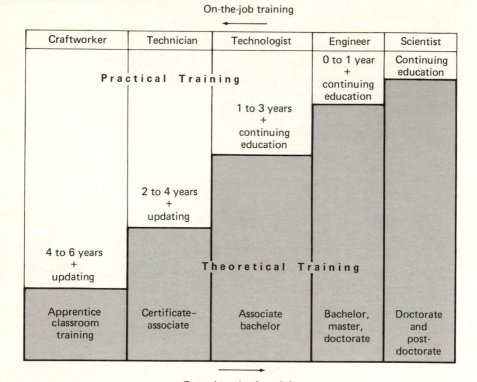

FIGURE 8-1
Education and training of members of the technical team.

THE TECHNICIAN

The engineering or physical science technician often uses the ideas of or carries out the technical plans of the engineer or scientist. Technicians are frequently the doers rather than either the innovators or the designers, even though they may do some drafting, design, or related work. An example is the electronics technician, who usually does one or more of the following: carries out the standard calculations for, estimates the cost of, or prepares service manuals for electronics equipment; installs, checks or tests, maintains and repairs, modifies and improves electronic equipment; sells or operates electronic equipment and/or facilities. Another example would be the drafter or the graphics specialist who converts an engineer's ideas and designs into detailed plans, working drawings, charts, graphs, schematics, and three-dimensional pictorials. Technicians are practical persons with an understanding of scientific principles, testing and measuring devices, and hands-on techniques. High school preparation may include either a general vocational or a college preparatory course of study. Normally this preparation includes drafting, elementary algebra, intermediate algebra, and physical science, but not an advanced mathematics or science course.

The surveyor is a specialist in the use of precision instruments in the measurement of dimensional relations among points, lines, and physical features. This level, set up on the downstream face of an earth-fill dam, permits the precise determination of differences in elevation. Because the surveyor's work is the basis for expensive construction and business negotiation, extreme accuracy and careful record keeping are imperative. (*Pacific Gas and Electric Company*)

The technician's education after high school typically requires two years of full-time study on the collegiate level. This is usually taken at a technical institute (or junior or community college technical program) which offers, in most cases, an associate degree. Associate degree courses normally include college algebra and trigonometry, report writing, considerable laboratory science, and training in equipment application and operation. Graduates of engineering technician programs are being accepted in increasing numbers for transfer into Bachelor of Engineering Technology collegiate programs with advanced standing, depending upon the individual's proficiency which is often demonstrated by examination.

THE TECHNOLOGIST

The engineering technologist is typically a practical person interested in applying engineering principles and in organizing people for industrial production, construction, or operation, or for improving devices, processes, methods, or procedures. Ordinarily one will deal with various components of an overall system designed and developed by engineers. A technologist frequently becomes a technical supervisor. In research and development one may be the liaison between scientist or engineer on the one hand, and technician and craftworker on the other hand. High school preparation for the technologist is often like that for the technician. Additional college education is usually pursued after receiving a two-year associate degree, and after two more years a Bachelor of Engineering Technology degree is awarded. This is frequently offered by a college of engineering or in a technical institute. Graduates of the two-year associate degree engineering technology programs usually will qualify for transfer into the third year of the bachelor of engineering technology programs. Since most of the general education, mathematics, and science courses are taken after the early concentration of technical courses (in contrast to engineering) they are sometimes referred to as "upside-down" programs. The educational emphasis of the technologist's program is less theoretical and less mathematical than that of the engineering counterpart but is more hardware and process-oriented. Typically, the technologist, as compared with the engineering technician, receives a broader education with more emphasis on general education, including business and supervisory training. Also, technologists often receive additional courses in technical disciplines and a greater depth of background in mathematics and science. The technologist's education, like that of the engineer and the scientist, should continue through lifetime.

THE ENGINEER

The engineer is primarily an innovator or designer of products, processes, procedures, or systems whose interest is in: how to solve optimally specific practical problems (through use of mathematics, knowledge of science, and practical judgment). In addition, the engineer is usually interested in how to solve problems economically. Typically, the engineer puts mathematics and

The operation of complex technological systems demands technical knowledge and good judgment. This load dispatcher controls the operation of a major segment in an electric power system. As the load starts to build up on a winter morning, he brings reserve equipment into standby condition. At the proper moment, he will connect in another generating unit or tie in another transmission line. For a given condition, he may have to choose among fossil-fuel, hydroelectric, or nuclear units on the basis of operating economy. (*Pacific Gas and Electric Company*)

A design engineer analyzes stress levels on a computerized model of a future front-wheel-drive car. Computer-aided design is one of the key factors in improving productivity. Computers not only speed design, but also facilitate product modification, and provide control and operation of production-line machines. (*Ford Motor Company*)

science to work in solving problems. But one may be involved in technical sales or become a technical manager. In leading technical organizations, engineers often become the top managers, chief executives, and presidents. The engineer's high school preparation in mathematics (preferably through advanced algebra, trigonometry, or a higher math) should be from above average to excellent. One should demonstrate an interest in and ability for reading scientific material, maintain average or good grades in English, have good study habits, possess good analytical and creative abilities, and understand people.

For most engineers the educational preparation after high school begins with either a 4- to 5-year full-time day engineering program, or a 5- to 6-year cooperative work-study program, or an 8- to 10-year evening educational program. Study and work experience should continue to expand their engineering background throughout their lifetime. Although most engineers do not pursue formal degrees beyond the bachelor's degree, a growing proportion of engineers are now seeking master's degrees in order to prepare to solve the latest technical problems; master's degrees in engineering are pursued either part-time while working in industry or government, half-time while employed in university teaching or research, or full-time as fellowships or assistantships. Some are being recognized by a professional degree (such as CE, EE, ME) in part for their contributions to engineering practice and in part for their advanced study. A number of potential engineering managers seek business administration or engineering management degrees. The doctor's degree in engineering and/or science is usually required for engineering teaching, advanced research, and development and research administration.

THE SCIENTIST

The scientist is typically the theoretician, the developer of hypotheses, the searcher for truth and the formulator of broad or basic ideas, and the discoverer and categorizer of knowledge whose interest lies chiefly in what occurs and why. High school preparation should include an above average performance in mathematics and the physical sciences, satisfactory work in English, and, if possible, a foreign language. Post-high school educational preparation begins with 7 to 8 years of full-time day (or longer part-time) undergraduate and graduate study—through the doctor's degree program. It continues throughout one's lifetime. One is expected to do research, to prepare articles for publication, to add to knowledge in the field, and to keep informed of new knowledge and theory. One's concern is usually with knowledge rather than its application or costs. Table 8-1 provides a comparison of various job positions for different team members related to the major branches of engineers.

Individuals who combine an interest in people, in economics, and in technical matters will find satisfying outlets for these interests primarily through engineering or technology, and sometimes through science. They frequently become sales, technical applications, or supervisory and managerial personnel. Some may manage their own businesses. Persons with sufficient educational background, an interest in the development of young people, and an interest in scientific technical ideas, may find satisfaction in formal teaching on the college level, in less formal teaching or training in industry.

Sometimes certain scientists, engineers, or other types of technical persons will be in short supply because society suddenly gives high priority to their particular field of competence. At such times, they may temporarily gain a financial advantage over their colleagues. However, the greatest long-term financial benefits tend to go to persons with integrity, technical competence, and good interpersonal skills who adapt readily to society's changing needs and priorities. To meet these needs, according to U.S. Department of Labor forecasts, greater and greater numbers of technical people will be required. Thus, as the population of the United States grows and as the nation's society becomes more complex, increasingly attractive opportunities will develop for skilled craftworkers, for well-prepared technicians, for

TABLE 8-1

REPRESENTATIVE TECHNICAL TEAM POSITION RELATED TO VARIOUS
ENGINEERING BRANCHES

Branch	Scientist	Engineer	Technologist	Technician	Craftworker
Aerospace	Aero-dynamicist	Aerospace design engineer	Airframe production foreman	Helicopter maintenance technician	Aviation power-plant mechanic
Agriculture	Agronomist	Agricultural engineer	Irrigation land surveyor	Foods technician	Farm-machinery mechanic
Chemical	Organic chemist	Chemical design engineer	Chemical laboratory technologist	Refinery technician	Oil-field equipment operator
Civil	Geologist	Chief sanitary engineer	Building construction contractor	Photo-grammetist	Carpenter
Computer	Scientist	Computer systems engineer	Computer programmer	Computer operator	Component tester
Electrical	Solid-state physicist	Electrical power engineer	Electronic equipment specialist	Electronic technician	Electrician
Industrial	Quality control statistician	Plant layout engineer	Production supervisor	Time and motion study clerk	Tool and die maker
Mechanical	Thermo-dynamicist	Mechanical design engineer	Manufacturing engineering technologist	Fluid power technician	Automobile mechanic
Nuclear	Nuclear research physicist	Reactor design engineer	Nuclear radiology specialist	Atomic power plant operator	Cooling system pipefitter
Materials	Physical metal-lurgist	Materials science engineer	Forge plant supervisor	Electron microscope technician	Electrical foundry operator

capable technologists, and for well-qualified engineers and scientists. The
opportunities will be there for those who are prepared.

EARNINGS

In general, physical scientists, technologists, and engineers receive higher
salaries than technicians and craftworkers but this is offset to some degree by

the longer education and training period required which not only entails expenditures for tuition but also loss of income. Figure 8-2 presents some comparative salary data for scientists, engineers, technologists, and technicians by years of experience. Where scientists and engineers with similar qualifi-

This solid-state physicist sees the first application of his research in a novel microelectronic circuit developed for improving radio communications under conditions hampered by extreme electromagnetic "noise" or interference. A primary application for the device, known as a surface-charge correlator, will be in jam-resistant radio communications systems developed for military applications. The wafer contains nearly 200 of the tiny circuits created by a technical team of physicists, chemists, electrical engineers, process technologists, and fabrication technicians. The novel circuit (a portion of which is visible on the television screen) was the subject of a technical paper presented recently at the Institute of Electrical and Electronics Engineers International Solid-State Circuits Conference in San Francisco. (*General Electric*)

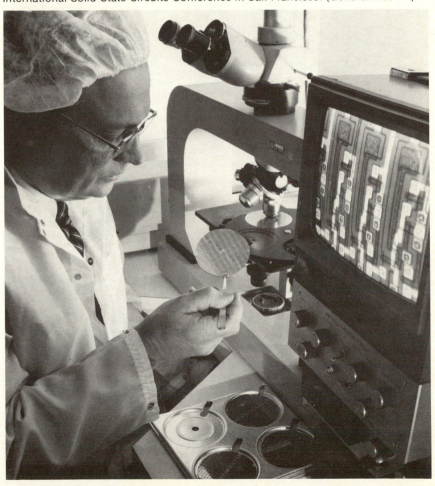

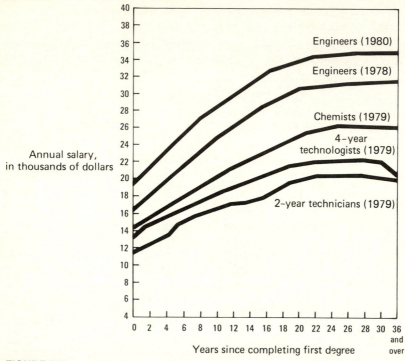

FIGURE 8-2
Comparative salaries of engineers, scientists, technologists, and technicians.
(Engineering Manpower Commission, *Professional Income of Engineers*, American Association of Engineering Societies, New York, 1980; *Salaries of Engineering Technicians and Technologists, 1979*, Battelle Columbus Laboratories, National Survey of Compensation, 1979.

cations are working in the same type of organization—for example, physicists and aerospace engineers in industrial research and development—their
salary scales are more likely to be comparable. The opportunities for engineers to move into high-salaried management positions give them an advantage at the upper end of the salary range. Some industries, however, are
beginning to recognize the value of the unique contributions of the technical
specialists by setting up salary schedules that provide the same compensation for the theoretical specialist with complex responsibilities as for the
technical manager with broad responsibilities.

While engineering technologists and technicians have relatively lower
salaries, their median salaries may peak earlier than for engineers and scientists. In recent years engineers' starting salaries have consistently been
higher than the starting salaries of college graduates in general.

WOMEN AND MINORITIES ON THE TECHNICAL TEAM

In the 1970s, national concern regarding affirmative action, equal educational and employment opportunities, provided the impetus for the engineering profession to examine more closely its constituents. Women and some minority groups were grossly underrepresented. For example, in the early 1970s, women and black Americans, Hispanic, and American Indian minorities constituted less than one percent of the total engineering profession, yet they represented 51 percent and 18 percent, respectively, of the national population. Some minority groups, however, including Asian Americans, have generally been well-represented and have made significant contributions to engineering. Some black Americans have made important contributions to engineering, including Lewis Howard Latimer, who worked with Thomas Edison and Alexander Graham Bell, and Fredrick M. Jones who held over sixty patents, including forty patents in the refrigeration equipment field [2].

Women engineers and scientists who have made significant contributions include Lillian Gilbreth who worked with her husband in industrial engineering on time and motion study; Edith Clarke who worked on power and circuit analysis in electrical engineering; Ellen Richards who worked in environmental and laboratory engineering; Kate Gleason who worked on worm gear design in mechanical engineering (and who became the president of a national bank); Elsie Eaves who contributed to engineering economics and civil engineering; Ivy Parker who worked on corrosion in petroleum and chemical engineering; Jenny Rosenthal Bramley who worked on electroluminescence and lasers in electrical engineering; and Betsey Ancker-Johnson who worked on plasmas in electrical engineering and formulated public policy as assistant secretary of commerce and was vice president of General Motors [3].

As part of a concentrated national effort involving industry and government working together with educational institutions, a significant program has been mounted to attract and retain women and underrepresented minorities in the engineering profession. Two of the leading groups in these efforts have been the Society of Women Engineers (SWE) and the National Action Council for Minorities in Engineering (NACME). In the latter half of the 1970s, engineering enrollments for women increased tenfold and for underrepresented minorities they increased fivefold. Although achieving parity or close to parity within the engineering profession and related scientific and technical occupations may be difficult to achieve, there are many indications that significant progress is being made and accelerated. Not only have freshman and undergraduate enrollments increased, but the number of degrees and graduate enrollments have also increased. Women and underrepresented minorities are also assuming increased responsibilities in industry, government, and education. Starting salaries and job offers for

These quality engineers reconstruct insulation material removed from the surface of the space shuttle's external fuel tank, seeking the reason for debonding of portions of the protective covering when cryogenic fuel and oxidizer were pumped into the tank. The engineers are part of a special team brought to the Kennedy Space Center to conduct the analysis. (*NASA*)

graduating engineering women and minorities have been equal to or greater than those for male and majority students. Although much remains to be done to adequately represent women and minorities, engineering is frequently cited as a profession where equitable opportunities are far greater than those provided in most other professional fields.

SUMMARY

The craftworker, the technician, the technologist, the engineer, and the scientist form an efficient team that is responsible for our technical progress.

There are many similarities among the members of the team but the qualifications and training required vary greatly.

REFERENCES

1 Johnson, A. Pemberton, *The Engineering Team,* Guidance Task Force, Engineers Council for Professional Development, New York, 1979.
2 Minorities in Engineering, "Black Americans in Science and Engineering," Afro-American Press, Indianapolis, Indiana, 1978.
3 Trescott, Martha M., "Women in the Intellectual Development of Engineering: Studies in Persistence and Holism," in Gabriel Kass and Simon and Patricia Farnes (eds.), *Intellectual History of Women in Science,* Stanford University Press, 1982.
4 "Professional Income of Engineers," American Association of Engineering Societies, New York, 1980.
5 "Interviewing and Employment Concerns of Women in Engineering," Society of Women Engineers, Purdue Chapter, West Lafayette, Ind., 1981.
6 Engineering Manpower Commission, *Women in Engineering,* American Association of Engineering Societies, New York, 1981.
7 Rose, Claire, et al., *The Study of Academic Employment and Graduate Enrollment Patterns and Trends of Women in Science and Engineering. Summary Report,* National Science Foundation, Washington, D.C., 1978.
8 Smith, J. Stanford, "Minorities in Engineering: A Five-Year Progress Report," *Engineering Education,* November, 1977.
9 Jagacinski, Carolyn and William K. LeBold, "A Comparison of Men and Women Undergraduate and Professional Engineers," *Annals of Engineering Education,* December, 1981.
10 Thomas Alva Edison Foundation, *Lewis Howard Latimer, A Black Inventor: A Biography and Related Experiments You Can Do,* General Electric Co., Greenfield Village, Michigan, 1973.

ASSIGNMENTS

1 Compare your attributes with those required for success as a technician, technologist, scientist, or engineer. Draw a conclusion regarding your future career.
2 Select a common article, and familiarize yourself with various phases of its manufacture and use. Describe, in a brief report, the work that would be done in connection with the selected article by a scientist, an engineer, and a technician. Devote a short paragraph to each.
3 Select a project, or an activity, and describe in general terms the differences in the work and responsibility of a technician and an engineer working together on the project.
4 Select a development project and describe in general terms the differences in the activities and responsibilities of an engineer and a scientist working together on the project.
5 (a) List five specific positions that would be appropriate for a technician and for

an engineer in a computer manufacturing company and describe, briefly, the work done in each. (*b*) Repeat for a spacecraft launch facility. (*c*) Repeat for a plastics processing plant.

6 (*a*) List five specific positions that would be appropriate for a scientist and for an engineer in a food packing plant and describe, briefly, the work done in each. (*b*) Repeat for a computer manufacturing company. (*c*) Repeat for the planning office of a large city.

7 Compare the monthly salaries earned in your community (in a single industry if possible) by a typical technician, technologist, engineer, and scientist. Obtain data for 27-year-olds (the scientist would be just starting work) and for 40-year-olds. Information may be obtained from personal interviews, employment managers, employment agencies, or publications. Analyze your findings, and draw a brief conclusion.

8 In a certain organization, the group leader had made it clear that each engineer and scientist is expected to publish at least one technical paper a year. Their new manager was surprised to discover how much of the group's time was being spent on publications. In an attempt to increase the productivity of the group, the workers are directed that no company time be devoted to publishing. Discuss the old and new policies and their effect on design engineers and research scientists.

9 What specific aspects of engineering should appeal to men, women, minorities, and handicapped persons? What problems does the woman engineer face in attempting to fulfill her traditional role while she pursues a professional career? How can students and practicing engineers make it easy for underrepresented minorities to be assimilated into study or work groups?

10 Interview an engineer, technologist, or scientist (or read about one in the library), and describe his or her background and current activities. Discuss any obstacles he or she has had to overcome and any special satisfactions he or she has received from work.

9

STATISTICAL PICTURE OF THE ENGINEERING PROFESSION

Is engineering an expanding profession? Where do most engineers work? For whom do they work? In which branches and functions are the largest numbers of engineers employed? How much do engineers earn, and how is their income affected by experience and education? Briefly, what are the facts about engineering?

This chapter provides a statistical description of the engineering profession as revealed by surveys made by various organizations in the past 10 years. Unfortunately no one agency has ever studied all aspects of the entire profession at any one time. To bring out the picture, we shall have to piece together bits of information obtained from different sources having slightly different viewpoints.

Those aspects of engineering that interest beginning students will be considered here; a more complete picture can be obtained from a detailed study of the references cited at the end of this chapter. The growth of engineering; the distribution within the profession by branch, function, education, and type of employer; the earnings of engineers as influenced by various factors; and the outlook for future employment in engineering have been selected as topics for discussion.

207

GROWTH OF THE ENGINEERING PROFESSION

Engineering is a dynamic occupation that is growing at a variable rate (Fig. 9-1). As indicated in Figure 9-1, prior to the early 1960s the total number of engineers increased more rapidly than the total population, but at a rate more comparable to the growth of the gross national product [1,2]. Gross national product (GNP) is the market value of the final goods and services produced by the nation's land, labor, and capital resources. However, since the mid-1960s, engineering growth has more closely followed the U.S. population growth.

The relative growth of engineering is clearly revealed in Figure 9-2. Prior to 1970, the number of engineers increased much more rapidly than the civilian and industrial work force. However, in the past decade as the United States economy has shifted from primarily production of goods to services,

FIGURE 9.1
Growth in the number of engineers in relation to total population and gross national product [5].

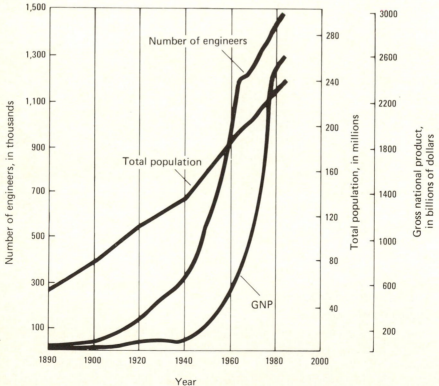

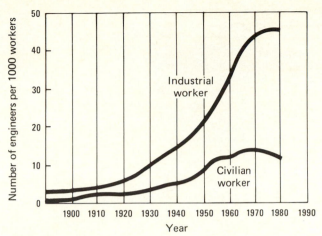

FIGURE 9.2
Number of engineers per 1000 workers. (U.S. Bureau of the
Census, "Statistical Abstract of the United States, 1981," 102d
edition; "Historical Statistics of the United States, Colonial
Times to 1970," 1976.)

the number of engineers per civilian worker has declined, but has continued
to grow at about the same rate as industrial worker growth.

In the past 50 years, the civilian labor force and population have about
doubled, but the number of engineers has increased sixfold so that the
number of civilian workers per engineer dropped sharply. In recent years,
however, the number of industrial workers and engineers has not grown
relative to the total civilian labor force. The increasing growth and complexity of science and technology and the problems related to growth including
environmental concerns and energy shortages have created increasing demands for engineers not only in the United States, but throughout the world.
As a result, in the early 1980s there have been unprecedented demands for
engineers and a parallel growth in the number of students who are pursuing
engineering and engineering technology careers in United States postsecondary educational institutions.

Geographical Distribution

As the engineering profession grew, its geographical distribution shifted. Engineering activity, along with industrial and commercial activity, tends to be
concentrated in population centers. Exceptions include mining in the
remote areas of the Mountain region and petroleum engineering in the
West South Central region. Figure 9-3 indicates the geographical distribu-

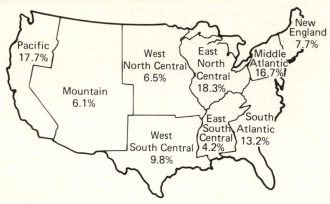

FIGURE 9.3
Geographical distribution of engineers. [2]

tion of United States engineers in 1978. Although similar to the regional distribution of the United States population, the larger numbers of engineers are employed in the more populous and industrial states. About half of all United States engineers are employed in eight states: California, New York, Texas, Pennsylvania, Ohio, Illinois, Michigan, and Massachusetts. After World War II, there was a significant shift from the Northeast and North Central regions to the more rapidly growing South and Pacific regions, but there were only minor shifts in the 1970s.

Distribution by Branch

As indicated in Table 9-1, electrical and mechanical engineering are the largest branches. Civil engineering, which was the largest branch in 1950, is now third in size. The fraction of the total engaged in aerospace engineering doubled in the 1950s and then declined. The 1970s were characterized by an increase in the number of interdisciplinary branches including biomedical, computer, and environmental engineering.

COMPOSITION OF THE ENGINEERING PROFESSION

The available reports on the engineering profession vary considerably because of differences in composition of the groups surveyed. Census figures are valuable because of their nationwide scope, but self-classification probably results in some misclassification. Some of the most valuable studies were those supported and published in the 1970s by the National Science Foundation. Data were gathered from samples of those classified as en-

TABLE 9-1
TRENDS IN THE DISTRIBUTION OF
ENGINEERS BY BRANCH [1, 10]

Branch	1950	1960	1972	1978
Aerospace	3%	6%	4%	3%
Chemical	6	5	5	5
Civil	24	18	14	13
Electrical	20	21	22	17
Industrial	8	11	10	10
Materials	6	6	5	5
Mechanical	21	18	19	17
Other	12	15	21	20

gineers in the 1970 census. These post-censal surveys of experienced engineers were supplemented by periodic surveys of recent engineering and science graduates. Together, these surveys constitute significant substantial statistical data based on a scientific representative sampling that permits broad generalizations regarding the United States engineering work force [2, 3, 4]. Also pertinent are the studies conducted by the National Society of Professional Engineers [5], the College Placement Council [6], and the Engineering Manpower Commission [7, 8, 9].

Demographic Data

The estimated 1.5 million engineers in the United States have a wide diversity in their demographic characteristics. They live and practice in urban Chicago, suburban Scarsdale, small towns in rural areas in Kansas, and on remote Pacific islands. They are corporate presidents, design engineers, troubleshooters, whistle blowers, and frustrated computer programmers. They are astronauts completing vigorous physical training, model railroad buffs, joggers, and gourmets. They are single, married, divorced, and widowed. Yet, there are some dominant features and distributions that make it possible to describe and profile the engineer.

Although the typical United States engineer is a native born, 42-year-old, married white male, with a bachelor's degree in engineering, there is an increasing number of women, minorities, single, advanced degree, and non-U.S. citizens comprising the engineering profession. Prior to the 1970s, less than one-half of 1 percent of the engineers were women or Black Americans. In 1978, over 2 percent of the engineers were women and over 1 percent were Black Americans. In 1981, 10 percent of the B.S. degrees awarded in engineering were to women, and 2.5 percent were to Black

Americans. In 1981, 17 percent of the first year engineering students were women and 6 percent were Black Americans. These data suggest that engineering is becoming more diverse and more representative of the population as a whole.

Distribution by Educational Level

Although the bachelor's degree is the dominant degree in engineering, there has been significant growth in the number and proportion of engineers with advanced degrees.

In 1978, about 4 percent of the engineering work force had doctor's degrees, 20 percent had master's degrees, 72 percent had bachelor's degrees, and about 5 percent had less than a B.S. degree. However, about 9 percent of those engineers 35 to 39 years old had doctor's degrees and 31 percent had master's degrees. In contrast, of those 55 to 59 years old, 3 percent had doctor's degrees, 17 percent had master's degrees [12]. These data suggest that advanced degrees in engineering are becoming increasingly common; however the B.S. degree will continue to be the dominant degree of experienced engineers in the next decade.

The educational level needed for various types of engineering employment is somewhat dependent on the type of industry or employer, branch of engineering, and functional responsibility. Table 9-2 summarizes the educational level by the branch of engineering.

Engineers involved in research and development functions, employed in high technology industries, and involved in higher education and majoring in aerospace, chemical, and nuclear engineering are more likely to have advanced degrees. Whereas engineers involved in operations, production, and sales-service functions, employed in established industries, in construction, and in utilities, and majoring in mechanical or civil engineering are more likely to have the B.S. degree [10].

TABLE 9-2
HIGHEST DEGREE BY ENGINEERING BRANCH [1]

	Doctor's	Master's	Bachelor's	Other	Total number
Aeronautical	4.5%	38.0%	55.2%	2.5%	48,700
Chemical	11.5	21.1	63.6	3.8	67,900
Civil	2.2	14.4	76.6	6.8	183,500
Electrical	4.8	20.7	67.1	7.4	243,000
Mechanical	2.4	18.7	72.9	6.1	233,700
Other	2.9	19.6	75.1	2.4	619,500
Total	3.5	19.7	72.3	4.6	1,396,400

Although the proportion of United States bachelor degree recipients in the 1970s who pursued full-time graduate work immediately upon graduation remained relatively constant (10 to 15 percent), an increasing number of foreigners have been pursuing graduate engineering study in the United States.

In addition, the strong quantitative and scientific base of an engineering education has resulted in significant efforts to attract engineers into other advanced degree disciplines including business administration, law, medicine, physical science, computer science and other quantitatively based disciplines. Many engineering graduates have also pursued advanced degrees in engineering and nonengineering areas (especially MBA programs) on a part-time basis. As a result, the majority of the recent recipients of B.S. degrees in engineering are pursuing advanced degrees or some form of continuing education within 5 years of graduation.

Distribution by Industry

Engineers are widely distributed throughout the work force, but their major concentration is in private industry, especially manufacturing. About three-fourths of the total are employed in private industry as may be noted in Table 9-3. About one-sixth of the total are in the electrical and computer industry.

A more detailed analysis of type of employer by branch of engineering can be found in the NSF references at the end of this chapter. These reports reveal that aeronautical engineers are likely to be employed in transportation especially in the aircraft and space vehicles; chemical engineers in chemical, petroleum and rubber industries; civil engineers in construction and engineering architectural firms; electrical engineers in electrical, electronic, and computer manufacturing as well as electric and communications utilities; mechanical engineers in machinery, motor vehicles, and manufacturing; industrial engineers in a wide diversity of manufacturing industries; metallurgical engineers in the primary metals and other manufacturing industries; mining engineers in coal, metal, and other mineral extraction fields; and petroleum engineers in the crude oil, natural gas, and extraction industries.

Distribution by Function

Engineers have a wide variety of functional responsibilities including research, development, design, production, construction, operations, sales, and management. The NSF surveys of experienced and recent engineering graduates provide some interesting insights on primary work activities. Research, development, and design are more apt to be the primary functional work activities of recent graduates, especially those with M.S.

TABLE 9-3
ENGINEERING EMPLOYMENT BY TYPE OF EMPLOYER [3, 4]

	Number (in thousands)	Percent
Business and private industry	1059	77.8
Manufacturing: private industry	567	41.7
Primary and fabricated metals	51	3.7
Machinery	110	8.1
Electrical; including computers	146	17.7
Transportation	115	8.4
Instruments	55	4.0
Chemicals, petroleum refining	50	3.7
Other goods	41	3.0
Nonmanufacturing: business	257	18.9
Petroleum, gas, mining, etc.	25	1.8
Construction	49	3.6
Business, financial, and real estate	62	4.6
Miscellaneous and other services	113	8.3
Utilities-communication, electric, gas, etc.	235	17.3
Education, colleges, universities, and other	52	3.8
Government	203	14.9
Federal-civilian and military	109	8.0
State and local	57	4.2
Other government	37	2.7
Other Nonprofit organizations	47	3.5
Total	1361	100.0

degrees. Management on the other hand is more apt to be the primary responsibility and work activity of experienced engineers [2, 3].

The 1981 survey of members of the National Society of Professional Engineers (NSPE) provides a more detailed functional breakdown by engineering experience and branch (Table 9-4). Note that design, production, and construction are fairly common initial functional responsibilities for less experienced engineers, whereas participation in management functions increases with experience. Aeronautical, agricultural, chemical, metallurgical-materials, and nuclear engineers are more likely to be involved in research and development than other engineers; aeronautical, electrical, and mechanical engineers are frequently engaged in design. Production is most likely to be the functional responsibility of industrial engineers whereas construction is

TABLE 9-4
FUNCTIONS OF PROFESSIONAL ENGINEERS BY EXPERIENCE AND BRANCH [5]

Years of experi- ence	No.	R&D	De- sign	Pro- duction	Con- struction	Sales	Manage- ment	Con- sulting	Teach- ing
0–4	1394	8%	35%	14%	12%	3%	6%	22%	0%
5–9	2673	5	26	8	13	2	15	29	1
10–14	2282	6	20	6	10	3	26	27	2
15–19	1956	5	15	6	8	3	34	25	3
25–29	2196	6	13	5	6	4	41	22	3
30+	3804	5	11	4	5	3	44	23	4
Total	16,488	6%	18%	6%	8%	3%	32%	25%	3%

Branches	No.	R&D	De- sign	Pro- duction	Con- struction	Sales	Manage- ment	Con- sulting	Teach- ing
Aeronautical	192	36%	23%	6%	2%	3%	18%	7%	5%
Agricultural	159	18	17	1	6	3	24	11	19
Chemical	5591	16	18	15	3	4	27	12	4
Civil	6286	1	18	2	12	2	34	29	2
Electrical	2655	8	25	8	6	5	28	18	3
Industrial	321	3	8	20	4	3	42	16	4
Mechanical	2978	9	23	8	6	4	25	22	2
Metallurgical– materials	86	23	3	26	0	1	20	20	7
Nuclear	138	16	9	12	7	1	32	16	7
Petroleum mining	377	3	9	15	4	3	41	22	2
Sanitary	1072	2	11	3	3	3	32	44	3
Total	14,855	6	19	6	8	3	30	25	3

most likely to be the responsibility of civil engineers. Consulting on the other hand is performed in all engineering fields and at all experience levels, but that may be due in part to the type of membership in the NSPE, which is generally limited to registered professional engineers.

Effect of Experience on Responsibility Level

Engineers take on a wide variety of technical and supervisory responsibilities. With increasing experience, engineers are given not only more technical responsibility but more supervisory responsibility as well. As noted in Figure 9-4, management functions characterize the jobs of many engineers with extensive experience. Table 9-5 provides data from the NSPE survey of engineers which relates both professional and managerial responsibility to length of experience. These data indicate that engineers have limited professional and managerial responsibilities during their first 5 years of employment. However, professional engineers take on rapidly increasing responsibilities involving not only complex and highly technical responsibilities but also increasing supervisory and managerial responsibilities. Although only a few become national/international authorities or assume major supervisory-managerial responsibilities for a large number of employees (500 or more), the majority of those with 10 or more years of experience have fairly challenging and complex technical responsibilities, and substantial supervisory responsibilities as well.

FIGURE 9.4
Increasing professional and managerial responsibility with experience [5].

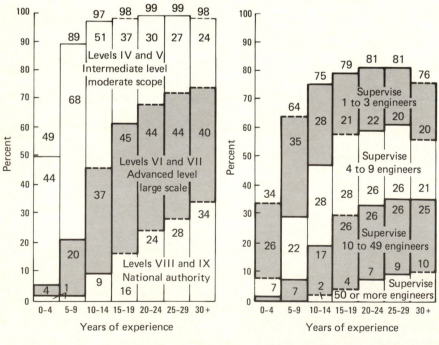

(a) Professional responsibility   *(b)* Managerial responsibility

TABLE 9-5
PROFESSIONAL AND MANAGERIAL RESPONSIBILITIES OF PROFESSIONAL ENGINEERS

Professional responsibilities	Total	Length of experience, in years						
		0–4	5–9	10–14	15–19	20–24	25–29	30+
Prescribed and standard practices	2%	19%	1%	1%	1%	1%	1%	1%
Specific and limited assignments	7	40	13	2	2	1	1	1
Detailed and moderate-scope duties	18	31	40	21	13	10	9	8
Complex planning duties	22	8	29	32	26	21	12	15
Large scale advanced level duties	19	2	12	23	26	24	23	26
Organizational leader-authority	14	1	4	13	18	21	21	26
Exceptionally difficult extensive in scope	11	0	2	5	12	16	16	20
National-international authority	7	0	1	2	5	7	11	16
Managerial responsibilities								
None	22	39	30	20	16	14	14	19
Supervise 1–3 engineers	23	27	35	28	21	20	17	15
Supervise 4–9 engineers	27	13	25	31	31	30	27	26
Supervise 10–49 engineers	19	2	9	17	23	24	25	23
Supervise 50 or more engineers	10	1	2	5	8	12	16	18

Source: "Income and Salary Survey, 1979," National Society of Professional Engineers, Washington, D.C., 1980.

Distribution by Areas of Critical National Interest

Over two-thirds of the experienced engineering work force in 1978 were involved in areas identified by the National Science Foundation to be of critical national interest including national defense, energy and fuels, environment and pollution, space, and housing and community development. There are also thousands, but relatively small percentages, of engineers involved in health, education, crime prevention, and food and mineral resources. Table 9-6 summarizes the involvement of experienced engineers in critical national interest areas by major branch. Aeronautical engineers are primarily involved in national defense and space; chemical engineers in energy-fuel and environment-pollution areas; civil engineers in environmental-pollution and housing community development; electrical engineers in national defense, space, and energy-fuel areas; mechanical engi-

TABLE 9-6

ENGINEERS ENGAGED IN AREAS OF CRITICAL NATIONAL INTEREST [2]

Areas of critical national interest	All branches	Branch of Engineering				
		AAE	ChE	CE	EE	ME
National defense	18%	48%	4%	4%	4%	18%
Energy and fuel	17	4	28	8	14	23
Environment-pollution	12	2	26	21	4	12
Space	4	27	1	...	6	4
Community-development	2	1	1	8	1	1
Housing	2	...	...	10	1	1
Health	2	...	2	1	1	1
Food	2	...	4	1	...	2
Other (crime, education)	10	5	7	22	8	7
Not applicable	31	13	28	27	34	31

neers in energy-fuel, national defense, space, and environmental-pollution areas. Although most engineers work in private industry, substantial support for engineers comes from federal agencies who are primary contractors for the federal government. Over 40 percent of the experienced engineers in 1978 reported they were working in fields supported by federal agencies. These data indicate that the supply and demand on the engineering work force are highly dependent on national priorities and federal policies [2].

EARNINGS OF ENGINEERS

For three decades, engineering graduates have received relatively high starting salaries. Their subsequent earnings are dependent on experience, the level of technical and supervisory responsibility, type of employer, branch and degree level, and the state of the economy. Salaries are also somewhat dependent on contemporary salary-wage policies, individual mobility, working conditions, and local and regional salary scales. Many of the intrinsic factors, such as challenge, productivity, creativity, technical and supervisory responsibility, and job commitment are dependent on the individual's initiative, interest, and aspiration level. However, there are other factors that are extrinsic and somewhat removed from the control of the engineer including organizational policies regarding salaries, fringe benefits, and promotions, as well as industry-company growth, labor-management relations, and research and development funding. In addition, national priorities, national defense, the economy, import-export policies, new discoveries, and technical obsolescence can also substantially influence engineering salaries and usually are not subject to the direct control of the individual engineer, the employer, or the engineering profession.

Starting Salaries

Table 9-7 provides information on starting salaries based on the annual surveys of the College Placement Council; the data for other years and

TABLE 9-7
MONTHLY SALARY OFFERS TO BACHELOR'S DEGREE
CANDIDATES [6]

Field of study	1974*		1981	
	Male	Female	Male	Female
Business				
Accounting	$925	$923	$1418	$1418
Business, general	809	756	1375	1315
Marketing	782	721	1293	1227
Engineering				
Aeronautical	960	994	1812	1840
Chemical	1042	1033	2031	2027
Civil	967	971	1771	1796
Electrical	986	1001	1882	1886
Industrial	978	1015	1839	1859
Mechanical	1001	1004	1907	1911
Metallurgical	1003	1015	1913	1921
Mining	NA	NA	1942	1929
Petroleum	NA	NA	2224	2206
Technology (engineering)	934	900	1809	1792
Humanities	728	655	1275	1157
Economics	NA	NA	1389	1336
Social sciences	766	696	1270	1099
Sciences				
Agricultural	789	728	1304	1206
Biological	751	664	1315	1222
Chemistry	891	867	1653	1612
Computer science	920	895	1736	1709
Health (MD professions)	727	734	1557	1305
Mathematics	878	871	1641	1607
Other physical sciences	898	878	1854	1813

*Data taken from 1974–75 pilot study.
NA—not available.

from other sources show similar trends. Bachelor of science engineering graduates consistently report receiving substantially higher starting salaries than B.S. graduates in most other fields. The differentials between male and female salaries in engineering are very small compared to most fields. Within engineering, chemical and petroleum engineers in recent years reported receiving higher starting salaries than those in other branches because of the high demands for energy, oil, and national defense. On the other hand, recent declines in construction may explain why civil engineers tend to report slightly lower starting salaries than those in other branches.

Figure 9-5 provides comparative data on starting salaries for new engineering and engineering technology graduates by degree level and the consumer price index data (CPI) from 1964 to 1981. Starting salaries of all engineering and technology graduates have kept pace with the CPI, with variations due in part to supply and demand and related economic conditions. Doctoral engineering graduates receive over twice the starting salaries of engineering technicians and about 50 percent more than B.S. engineers.

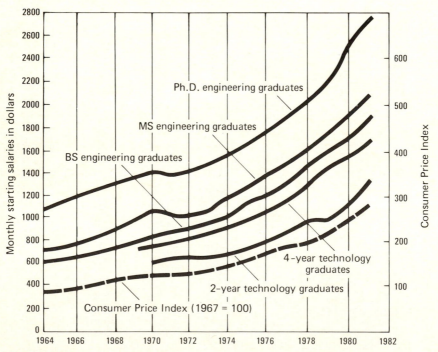

FIGURE 9.5
Starting salaries of technologists and engineers [6].

Bachelor of science engineers receive about 50 percent more than 2-year technicians and about 10 percent more than B.S. technology graduates, but about one-tenth less than M.S. engineering graduates. However, as may be noted in Table 9-5, all the engineering and engineering technology graduates received relatively high starting salaries compared to college graduates in other fields.

Professional Income

Engineering salaries are related to a wide number of factors including experience, degree level, technical and supervisory responsibility, geographical location, and type and size of employer. There is also considerable variation, even for engineers with similar jobs and similar experience. Among the

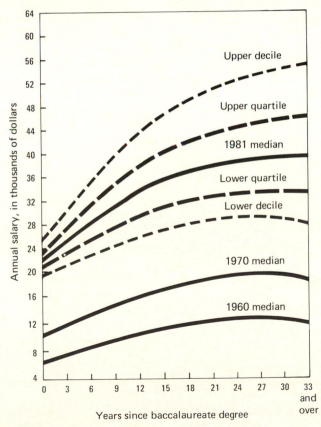

FIGURE 9.6
Income of professional engineers [7].

most comprehensive and up-to-date salary surveys conducted within the engineering profession are the biennial surveys of major engineering employers conducted by the Engineering Manpower Commission. Figure 9-6 provides data on annual salaries for engineers with varying levels of experience. By way of explanation, one-half of those surveyed earned more than the median, one-fourth earned more than the upper quartile, and one-tenth earned more than the upper decile.

Figure 9-7 provides median annual income data by degree level. The data provided some insight into the relative relationship of these variables. Those with a doctor's degree have the highest median earnings and those with B.S. degrees have the lowest salaries. However, caution should be taken in interpreting and generalizing these since for each degree level there is a family of upper and lower decile and quartile curves similar to those in Figure 9-5. Highly competent engineers with limited formal education earn more than average engineers with advanced degrees. Furthermore, those who pursue advanced degrees defer the start of full professional compensation.

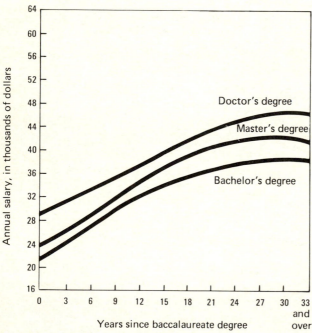

FIGURE 9.7
Median income of engineers by degree level [7].

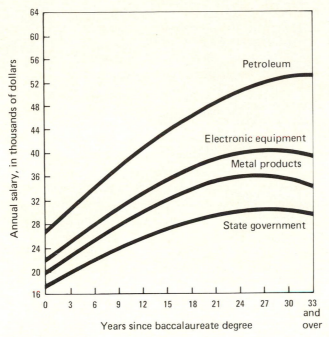

FIGURE 9-8
Median engineering salaries by industry [7].

Salaries are also dependent on the type of employer as well as experience. As may be noted in Figure 9-8, even the starting salaries for those employed in various industries and organizations differ significantly. Note that the differences in salaries become greater with experience.

NSPE Income and Salary Surveys

The NSPE has conducted biennial surveys of their members since 1952. These data tend to be consistent with those found by the Engineering Manpower Commission. They also provide detailed breakdowns by degree level, branch of engineering, function, and degree of managerial responsibility. Figure 9-9 provides a different view of engineering income because NSPE members have met registration requirements, and therefore this is a selected group. However, the effects of experience, clearly shown in Tables 9-4 and 9-5, are not obvious here.

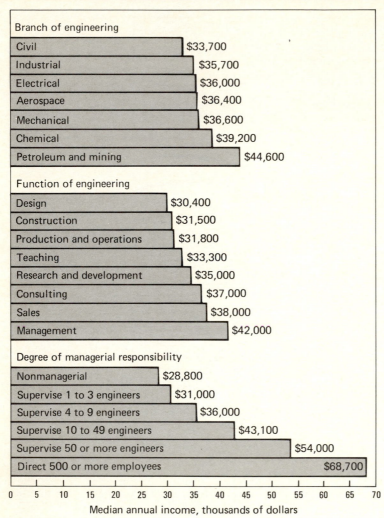

FIGURE 9.9
Median salaries of NSPE members [5].

EMPLOYMENT OUTLOOK FOR ENGINEERS

As the Bureau of Labor Statistics noted in their 1982–83 *Occupational Outlook Handbook:*

> Employment opportunities in engineering are expected to be good through the 1980s. In addition, there may be some opportunities for college graduates from related fields in certain engineering jobs.

Employment of engineers is expected to increase slightly faster than the average for all occupations through the 1980s. In addition to job openings created by growth, many openings are expected to result from the need to replace engineers who will die, retire, or transfer to management, sales, and other professional jobs [13].

The Bureau of Labor Statistics[1] examined the trends in employment and the expected supply and demand for the 1980s. In engineering they projected a 27 percent increase from 1980 to 1990 overall. They predicted an average annual demand of about 46,500 engineers, 25,500 due to expected growth and 21,000 to replace engineers due to retirement, death, changes in employment, etc. These somewhat conservative projections suggests that those interested and qualified to pursue engineering as a career can expect to find plenty of initial jobs with rapid advancement opportunities.

There exists the normal engineering growth caused by the ever-increasing U.S. and world demand for goods and services. But, in addition, there are new technologies—involving microcomputers, exotic materials, expanding communications media, and new energy sources—that can be expected to provide opportunities for well-qualified professional engineers. Increasing national and worldwide concern for defense and arms control, ecology and environment, and housing and hunger, may combine to create an unprecedented demand for professional engineers.

On the supply side, the increasing number of women, minorities, and foreign nationals who are preparing themselves for engineering careers should help. Information regarding improved career opportunities, especially in engineering, has resulted in increases in the number and proportion of college students who major in engineering. We have also observed that in this time of engineering shortages, foreign engineers and individuals in related fields tend to migrate into the U.S. engineering work force [14].

What makes short- and long-term engineering supply-and-demand estimates difficult is their dependency on the amount and distribution of federal expenditures for energy, defense, and research and development, as well as the industrial expenditures for new plants and equipment, research and development, and new ventures in goods and services [15].

Nevertheless, when compared to most career fields, engineering seems most promising for the next quarter of the century. Engineering graduates

[1]Bureau of Labor Statistics, "Occupational Projections and Training Data," 1980 ed., no. 2052, Washington, D.C., 1980, p. 88.

have had a higher proportion of job interviews or job offers, and higher start-ing salaries than any other baccalaureate field. Engineers have always had lower unemployment rates than the national average, not only among the total civilian work force, but also among the college educated work force including most of the professional and technical work forces. Long-term unemployment in engineering has been virtually nonexistent, and short-term unemployment has usually been well below national averages. The overall statistical picture for engineering employment promises to remain relatively stable and positive.

SUMMARY

Career opportunities in engineering for the next decade look quite optimis-tic. The increasing challenge and dynamic nature of our ever expanding technological society, coupled with the increasing diversity in the engineer-ing work force, promises to provide excellent prospects for stimulating and exciting lifetime careers to those who enter the engineering field.

REFERENCES

1 National Science Foundation, "U.S. Scientists and Engineers, 1978," NSF 80-304, Washington, D.C.

2 National Science Foundation, "Characteristics of Experienced Scientists and En-gineers, 1978," NSF 79-322, pp. 65–66.

3 National Science Foundation, "Employment Attributes of Recent Science and Engineering Graduates," NSF 80-325,.

4 National Science Foundation, "Scientists, Engineers, and Technicians in Private Industry, 1978–80, NSF 80–320.

5 National Society of Professional Engineers, *Income and Salary Survey, 1981,* Washington, D.C., 1981.

6 College Placement Council, *CPC Salary Survey: A Study of Beginning Offers,* Bethlehem, Pa., 1981.

7 Engineering Manpower Commission, *Professional Income of Engineers, 1981,* American Association of Engineering Societies, New York, 1981.

8 Engineering Manpower Commission, *The Placement of Engineering and Tech-nology Graduates,* American Association of Engineering Societies, New York, 1981.

9 Engineering Manpower Commission, *Engineering and Technology Degrees,* Engi-neers Joint Council, American Association of Engineering Societies, New York, 1981.

10 National Science Foundation, "The 1972 Scientist and Engineer Population Redefined," 75–313.

11 U.S. Department of Energy, *1979 National Survey of Compensation,* Office of In-dustrial Relations, 1979.

12 U.S. Department of Labor, Bureau of Labor Statistics, "National Survey of Profes-

sional, Administrative, Technical, and Clerical Pay," March 1980.

13 U.S. Bureau of Labor Statistics, *Occupational Outlook Handbook,* 1982–1983.

14 Vetter, Betty M., *Supply and Demand for Scientists and Engineers,* Scientific Manpower Commission, Washington, D.C., 1977, p. 45.

15 Engineering Manpower Commission, *The Demand for Engineers, 1980,* American Association of Engineering Societies, New York, 1981.

ASSIGNMENTS

1 Write a paragraph (100–200 words) describing and explaining the relations between the number of engineers and the GNP shown in Fig. 9-1.

2 What shifts in geographical distribution of engineers do you anticipate over the next 20 years? Explain in terms of new developments and their effect on specific branches on engineering.

3 In what branches of engineering does possession of an advanced degree seem to be most important? Least important? Why?

4 In what functions of engineering would you expect graduate work to be most important? Least important? Why?

5 Write a paragraph (100–200 words) describing the data shown graphically in Fig. 9-6. What conclusions can you draw?

6 From Bureau of Labor Statistics bulletins for your area, obtain data on the hourly pay of journeymen electricians or plumbers for the period covered by Fig. 9-5. Assuming 170 hours of work per month, calculate monthly salaries and plot (on a graph similar to Fig. 9-5) the monthly income of electricians or plumbers and compare them to starting salaries for engineers.

7 Figure 9-6 presents simultaneous data for many different engineers and not the salary histories of the same engineers over a period of time. On a graph similar to Fig. 9-6 and using data from Fig. 9-5, sketch an estimated salary history of an engineer with 32 years of experience in 1981 and earning a median salary that year. Repeat for men with 24 and 15 years- experience in 1981, respectively.

8 Analyze Table 9-2 on the basis of your knowledge of the engineering branches gained from Chapter 5. Explain the significant differences between the branches in terms of the historical age of the branch, the dominant functions, and the desirable educational qualifications.

10

LEARNING SKILLS

If you are getting top grades in all your classes and still have time for social and extracurricular activities, don't bother to read this chapter. But if your academic achievement is not what it should be, or if your achievement is satisfactory but you never seem to have any free time, you may wish to take advantage of what is known about efficient learning. Speaking frankly, most college students don't know much about effective studying even though they spend many hours a day "working at" it. It has been demonstrated that students with good records as well as those with poor records can benefit greatly from training in the proper techniques of study.

For the next four or more years you should consider yourself a "professional" student striving for efficiency in the management of the large investment in money, time, and effort that the college program represents. Having paid your fees and enrolled in a course, you are entitled to sit in the class and take the examinations for credit and a grade. But the class itself is just an indication of the educational opportunities available to you. All the information, facts, and principles presented in most college courses could be bought at the bookstore for a few dollars. The true measure of the educational value

228

of the course is what happens to you—how your attitudes are modified, how your viewpoint is enlarged, how your reasoning power is expanded, how your perception is deepened, how your ability to make decisions is developed.

Teaching is what the professor does but learning is what you do, and learning is a skill—an art—derived from a knowledge of principles and perfected by regular practice. For the skillful student, learning is a pleasure rather than a chore.

PRINCIPLES OF LEARNING

Psychological research reveals that learning is accomplished most readily when there is motivation, concentration, comprehension, action, distribution, and evaluation.

The large majority of freshmen (this will include you!) go through an adjustment when they start college, while they enjoy their new freedom from parental control. This adjustment lasts 2 to 3 weeks for most students, 2 to 3 months for some students, and is never completed by a few. The first phase of this adjustment occurs when students realize that doing homework regularly is necessary for success and that no one is going to make them do it. The next phase is designing a good study schedule. The final phase is making adjustments to this schedule, finding good places to study, and conscientiously following the study schedule. You will find that hours in the library or areas reserved for study are much more productive than the same time in a noisy dorm. "All-nighters" may be necessary once in a while but they are a poor substitute for regular homework.

Motivation

A motive is an inner drive that causes a person to act in a certain way. A hungry rat is highly motivated to solve a maze when it knows there is food at the end. "Hungry" students with a clear goal ahead and a strong desire to succeed look forward to a study session as another step in their career, and their receptive frame of mind makes studying satisfying and effective. Poorly motivated students have no particular interest in their courses, find it difficult to bring themselves to studying, and accomplish little even when they are at their desks with books open in front of them. In which category are you?

Motivation is more important than any other single factor in determining success in college. In contrast to high school, you are on your own and an inner drive is essential. For some students the desire to learn provides sufficient motivation. If you have a tentative vocational choice such as engineering, you have an advantage in that many of your courses have a relationship to your ultimate goal. As a rule, students who hold outside jobs during school earn slightly better grades than those who don't; perhaps their higher motivation more than compensates for the loss in time available for study. Lack of motivation is deadening. If after your study of the first part of this book you have no clear career goal, you should take advantage of your college's counseling services immediately. Until you have a purpose in mind and a desire to learn, your studying will be inefficient at best.

Concentration

The human mind is a wonderful thing. You can sit in a crowded room exchanging pleasantries with a companion while enjoying a familiar melody in the background, and still get the drift of the conversation behind you. Your mind can perform several functions at once if they are all routine, familiar, and of no great consequence. On the other hand, if you are trying to master a new concept in calculus or produce an original theme for your English class, you need to focus all your mental faculties on a single object.

The essential requirement for productive study is an eager, receptive mind; the properly motivated student can work under almost any conditions. For example, concentration on an absorbing novel is possible under conditions that otherwise would be distracting. However, because efficient study requires an all-out effort, all distractions, annoyances, or discomforts should be minimized and a perfect physical situation provided.

It isn't so much how *long* you study but how *well*. Two 30-minute periods of concentrated study will accomplish more than two hours of mixed study, relaxation, and daydreaming. Be sure to distinguish between concentration and relaxation; keep your study area sacred for studying (whether it is your room or the library); relax somewhere else. When you're sitting in that chair at that desk, study. When you feel the need to relax, stretch out on the bed or go down to the lounge. Getting your roommate and friends to cooperate may be difficult at first but it will be well worth the effort. Once the ability to concentrate is developed, you will be able to do better work in less time, and time previously spent in inefficient, half-hearted study will become available for other pursuits.

Comprehension

An important principle of learning is that before material can be mastered it must be understood. Compare, for difficulty in learning, a proverb in English with the same number of words in an unfamiliar language. The plot of a novel is easy to grasp but a new physics concept is much more difficult. Sometimes, learning the meaning of new words is essential for comprehension; in other cases a review of previous discussion or subsequent application will be necessary. An attempt to memorize the statement of a theorem without comprehension is useless. A good test of understanding a concept is to close the book and write out an equivalent statement in your own words.

Action

In spite of claims for under-the-pillow training devices, learning is not a passive process. The mind of the learner must be receptive in an eager, aggressive way. The defensive linebacker in football is a good example of physical alertness. He is striving to sense at the earliest instant which way the play is going to develop; poised on the balls of his feet with muscles slightly tensed, he is ready to jump in response to hints and clues. In the same way the effective student must be alert to each point made by the lecturer or each key phrase in the text.

Letting the lecturer's words pass through your mind or letting the printed page pass before your eyes accomplishes nothing. You must perform some mental act on each bit of knowledge to make it yours. Studying the page to prepare yourself to answer a question or solve a problem gives purpose to your reading. Comparing the author's statement to your own experience permits assimilation. Rephrasing the professor's statements into an outlined set of lecture notes involves the type of activity that contributes to learning. Usually it is sufficient to look over your class notes and add comments to insure comprehension. Normally it is not worth the time to recopy them.

Distribution

A given learning task is accomplished much more effectively if the learning is distributed over a period of time. Research on the rate of forgetting demonstrates that forgetting is considerably slower when short periods of study at regular intervals are used instead of a single prolonged effort. In college work this means regular daily studying is preferable to all-night cramming sessions before examinations.

Combinations of reading and reciting or reviewing and reciting are definitely more effective than straight studying. Recitation also provides a check on comprehension and, by focusing attention on unlearned portions of an assignment, encourages proper restudy for complete understanding. The optimum length of study sessions for long-term retention will vary with the material; in general, the sessions should be shorter and more frequent than is customary with most students.

Evaluation

The analogies between cultivating a trained mind and developing a trained body are clear and accurate. An important contributing factor to continually improving athletic performances is the readily available quantitative evaluation. With the 3:50-minute-mile barrier smashed, a 3:45-minute mile is beckoning; breaking through the 18-foot pole vault ceiling opens up new horizons. The stopwatch and the tape measure provide an accurate and ever-present evaluation of performance.

Psychologists tell us that we all do better when we have a method of comparing our achievements with those of others or a standard for evaluating our own efforts—in a sense competing against ourselves. We improve more when we see we are improving; the feeling of accomplishment is an effective incentive in itself. Your goal in improving the learning process is to accomplish more of a higher quality in a shorter time. Wherever possible, you should keep continuous records of quantitative measures such as examination grades, grade point average, or rate of reading. To a professional student these should be just as important as a professional baseball player's batting average. Where quantitative measures are not so readily available, it is well worth the effort to develop criteria for evaluating your techniques as you strive to improve the effectiveness of your study methods.

TIME BUDGET

Hold out your hand for a reading of your palm. On your palm is written that you are not getting as much studying done as you feel you should. You have difficulty getting down to work. You waste a lot of time because you do not feel like studying or because you cannot make up your mind what to do next. You are always behind in your work and never quite catch up. At the end of each week you wonder where all the time went and you are disappointed at how little you have accomplished.

Why was this so clear in your palm? Because you are a college student and practically all students feel this way to a greater or lesser degree. Students come to school with widely varying financial resources and with greatly differing mental resources; but everyone has the same time allowance—just 24 hours a day. To take advantage of all the educational opportunities that a college offers, to participate in other valuable activities, to earn some of your college expenses, and to still have time for eating and sleeping requires the same careful planning and scheduling that an executive gives to his or her business operations. It cannot be catch-as-catch-can or you will find yourself frittering away valuable hours, running out of time when you need it, working feverishly when you need some relaxation, and unable to relax when you put aside your studying. If you have difficulty organizing your time in an efficient way, a time budget may be very helpful.

Reasons for Scheduling Your Time

A financial budget is not for the purpose of saving money but rather for ensuring that money will be available for buying desired items. Similarly, a time budget will help to make sure that time is available for recreation, social activities, and sleep, as well as for study. One great advantage of following a schedule is that it provides additional time by eliminating the time wasted in deciding what to do next. A schedule turns odd hours that might easily slip away into solid achievement. Also it encourages efficiency by clearly designating what a given hour is to be used for. It is a poor policy to mix social activities with study sessions; on the other hand, there is no benefit from recreation time that is spent worrying about unfinished assignments. Properly planned, a time schedule puts the time you have available into the things you wish to do.[1]

A Typical Schedule

Figure 10.1 shows how a freshman engineering student might budget her time. Her class schedule is shown below. In addition to the 24 hours in class and laboratory and the 23 hours of outside study, she works 8 hours per week. After several trials and with some adjustments, she planned her 55-hour week as indicated on page 234.

---

[1] Some physiologists believe that persons vary in regard to their cycles of greatest mental activity. Some, the A.M. type, jump from bed wide awake and ready to work; usually they get sleepy relatively early and should not plan on late studying. The P.M. type, on the other hand, awake slowly, take a long time to get under way, but are wide awake at midnight. If possible, you should schedule important activities during your periods of greatest mental alertness.

FIGURE 10-1
A typical schedule.

Course	Units	Class hours	Outside hours	Type of assignment
Mathematics	3	3	8	Problems
English	3	3	6	Reading and writing
History	3	3	4	Reading
Chemistry	3	5	4	Principles and problems
Computers	2	6	0	Programming
Physical education	0	2	0	Exercise
Orientation	1	2	1	Reading, etc.
Total	15	24	23	

FIGURE 10-1

Time	Monday	Tuesday	Wednesday	Thursday	Friday	Saturday
7−8	*Breakfast*	*Breakfast*	*Breakfast*	*Breakfast*	*Breakfast*	*Breakfast*
8−9	**Math**	Work	**Math**	Work	**Math**	Work
9−10	**English**		**English**		**English**	
10−11	**Math**	**Orientation**	**Math**	**Orientation**		
11−12	**History**	**Chemistry**	**History**	**Chemistry**	**History**	
12−1	*Lunch*	*Lunch*	*Lunch*	*Lunch*	*Lunch*	*Lunch*
1−2	**English**	**Chemistry**	**Orientation**	**Chemistry**	**Math**	*Business and recreation*
2−3	**Computer lab**	*Personal*	**Computer lab**	**Chem lab**		
3−4		**Physical ed.**			**Physical ed.**	
4−5		**History**				
5−6	*Recreation*		*Recreation*	*Personal*	**English**	
6−7	*Dinner*	*Dinner*	*Dinner*	*Dinner*	*Dinner*	*Dinner*
7−8	**History**	**Math**	**English**	**Math**	*Makeup or recreation*	*Recreation*
8−9						
9−10	**Chemistry**					
10−11						
11−12						
12−7	*Sleep*	*Sleep*	*Sleep*	*Sleep*	*Sleep*	*Sleep*

A well-organized study area is a key part of your college environment. It is a comfortable feeling going to class with your homework finished and being prepared to ask questions on any material you didn't understand. Is this a regular part of your college activities?

To be effective, a time schedule must be tailored to suit you, your habits, and your course requirements. In the case of this student, she knows that she needs to get to sleep by midnight but that she has no trouble getting up early; she arranged for an 8 A.M. mathematics class three days a week and she scheduled work early on the other mornings. By taking full advantage of the daytime hours, she is able to finish her studying by 10 or 11 P.M.

Mathematics   There is a daily homework assignment in mathematics; since mathematics is important, difficult, and time consuming, the student plans on using eight hours per week for math work outside of class. On the same day as class, she spends an hour studying the new material and trying some of the easy problems to make sure she has grasped the new concepts. If not, the next day she sees her instructor during his office hours or checks with friends to get her difficulties straightened out. In this way she avoids the common trouble of struggling hour after hour on the night before an assignment is due because of some basic confusion or misunderstanding.

English   Usually an English paper is due each Monday morning. Either she has to do this writing on Sunday (which she tries to avoid) or she has to do it ahead of time because she also has mathematics to do for Monday. Each Wednesday evening she schedules a four-hour session in the library looking up references, compiling notes, and writing out a rough draft. Thursday morning before going to work she looks it over again and changes the portions which, inevitably, do not sound so good on the second reading. Friday afternoon she reads it again, makes minor modifications, types it up, and puts it away for Monday morning.

Chemistry   Our engineering student tries to spend an hour reading before the chemistry lecture because if she reads ahead in the text the demonstrations are much easier to follow and to understand and because the professor frequently springs a little 10-minute quiz on the previous lecture. Because the best time for review is immediately before recitation, she tries to get to chemistry class a few minutes ahead of time. The chemistry lab work can be finished in the allotted time if she keeps busy for the full 3 hours except for a couple of quick breaks.

Computer skills are a required part of any engineering curriculum. Here students learn to program and do their homework on remote terminals. As engineers, they will use more powerful computers to conduct experiments, design and display entire buildings, control manufacturing plants, optimize space-vehicle trajectories, or display three-color representations of microscopic integrated-circuit designs.

Computer Programming   Each computer session includes a short discussion of new material for which some text study may be required. The assignments consist of applying new programming techniques to various situations. By working industriously for 40 minutes, taking a 10-minute break for relaxation of mind and eyes, and then returning for another concentrated effort, a capable student can complete all the assignments during the computer lab.

Orientation   The one-unit freshman orientation class is primarily a lecture course with reading in the text and short assignments. Careful attention to the lectures and another hour per week of study is sufficient for the formal requirements. Late evening hours and weekends provide time for reading, bull sessions with other engineering students, and informal talks with practicing engineers.

Physical Activity   For the sake of good health, any time budget should make provision for some daily physical activity, if possible. This student takes a no-credit physical education class two days a week, and her job also involves some physical labor. These activities plus walking between classes and to and from the campus provide the necessary relief from the normally sedentary life of a college student.

Recreation   An engineering student carrying a normal load and working part-time will not have many free hours during the week for recreation, which is pleasant, relaxing, and essential for top efficiency during work periods. As her friends who are enrolled in less exacting majors head for the movies, our student has to wave "good-bye" and get to work. However, it is not as bad as it looks because she has scheduled full hour periods for relaxed meals, an hour before dinner on Monday and Wednesday (her heavy days), and she has left her weekends relatively free. If she prefers to put in three hours on, say, Sunday afternoon, she could take a weeknight off for a play or concert. Also, if she works efficiently during her evening sessions, there should be time for a quick trip to the student union before going to bed. There is time for study, work, and recreation, but it has to be carefully planned.

Personal Development   On every college campus there are numerous opportunities for the interested student to enrich his or her life outside of the program of formal courses. Because the hour after chem lab and before dinner is easily wasted, it is ideal for a culture hour—a time for reading a novel, stopping in at the art exhibit, or listening to recordings. In addition, she plans on an hour or two on Sunday for activities that are spiritually or intellectually stimulating.

Preparing Your Time Budget

No one else can make out a time schedule for you. What is best for you will not work for your roommate, even if he is taking the same classes. To develop a successful schedule, first analyze your present study habits and time allocations, then prepare a tentative working schedule and try it out, and finally prepare a revised schedule based on your experience.

Analysis of Present Operation   Write down your class schedule and any other commitments such as part-time employment. Analyze your courses in terms of the nature of the preparation required and estimate the number of hours of outside work per week. Bear in mind how fast you work and how rapidly you read, but remember that college students tend to overrate their own abilities. Add in a reasonable allowance for recreation and social activities, for meals, for transportation, and for personal development. Does the total show adequate time for sleep?

Preparation of Tentative Schedule   Make out a form similar to Figure 10.1 and enter your scheduled classes, laboratories, meals, sleeping hours, etc. In planning your study periods, consider the total number of hours and the best possible distribution. Place study periods close to class periods if possible—before a class involving recitation or after a lecture during which you have taken many notes. Do not waste hours between classes; make definite allocation of all hours from 8 A.M. to 6 P.M. For drill or memorizing, keep study sessions short for high learning efficiency; for major projects, provide several hours in a row with suitable breaks. Be sure to include time for physical activity, recreation, and personal development. Provide some flexibility by including occasional unscheduled hours.

Preparation of Revised Schedule   You have planned your work; now work your plan. Put your schedule in the front of your notebook so that you can refer to it at all times. For at least a week or two stick closely to your schedule; give your tentative plan a chance to demonstrate how organization can help. Assuming your first schedule is not completely unrealistic, avoid drastic revisions; instead, shift a few hours here and there to secure minor improvements as inadequacies appear. If your schedule breaks down at times, note the reasons and analyze this information before preparing the next revision. After a few trials you should arrive at a practical time plan that should work under most situations. Once your hours are scheduled you can consider how to get the most out of each hour of lecture or study.

LECTURES

It has been said that "a lecture is the process whereby the notes of the instructor become the notes of the student without passing through the minds of either." To some extent this may be true, but the lecture is still a primary source of instruction for college students and many unsuccessful students are in academic difficulty because of their inability to profit from lectures.

Reasons for Taking Notes

A good set of lecture notes is an invaluable study guide—a record of points emphasized by the lecturer, a stimulus for further thinking or additional reading, and an organized summary for later review. On some campuses skilled students in large lecture classes take notes that are then duplicated and sold. Purchase of a set of notes, however, precludes some of the major benefits of note taking. One ingredient of successful learning is activity. Forcing yourself to comprehend the important points of a lecture, to organize them in your mind, and to then record them in meaningful form constitutes an effective type of learning activity. Good notes represent an assimilation of ideas by the note taker and not just an incomplete shorthand account of what was said. In addition, the attention note taking demands prevents your mind from wandering by keeping it occupied with questions such as: What is the point of what the lecturer is saying? Is this important? How is it related to what was said before? Of great importance is the fact that note taking from lecturers, readings, telephone conversations, inspection trips, etc., is a valuable skill that will have continual application throughout your professional career.

Note-taking Techniques

Here are some specific suggestions that work for the majority of students. Try them out, but feel free to develop your own methods wherever you think you can obtain better results.

1 *Obtain one good quality, three-ring, loose-leaf notebook for standard 8¹/₂- by 11-inch paper.* Using one notebook, you are always sure to have it with you. Your notebook will get heavy usage and it should be durable. Looseleaf notebooks permit rearranging material and inserting material handed out in class, most of which will be on 8¹/₂- by 11-inch paper.

2 *Organize your notebook for action and use.* Use labeled tabular dividers to separate your notes by classes. Keep a good supply of blank paper in the notebook. Date, number, and title each page for ready reference. Write on only one side of the paper. Write in ink for durability and legibility. Keep your notebook neat and avoid doodling; starting to doodle is a sure sign that your mind is wandering. Don't crowd; leave room for adding notes in the margin or on the back of the previous page while reviewing.

3 *Strive for useful notes.* To be useful, notes must be understandable several weeks later. Be brief but be sure that ideas are complete. If you are to depend on your notes, they must be accurate. Develop your own set of abbreviations to cover the words and phrases commonly used in your courses; in engineering you can take advantage of mathematical symbols, standard abbreviations, and Greek letters. Be sure to label all sketches.

4 *Organize your notes as you take them.* In some classes the instructor's organization is too vague or the ideas come too rapidly to permit taking notes in good form. Sometimes thinking through ideas, correlating points, and rewriting notes is an effective study procedure. In general, however, rewriting notes is time-consuming and the time is better spent on studying. An outline form (see page 259) is useful because it shows the main headings and subordinate points, and their relations to each other. Once you have learned an instructor's idiosyncrasies and can recognize his or her pattern of presentation, outlining is much easier. Listen for key terms such as "most significant properties," "basic principles," or "major applications"; these can be translated directly into main headings with subordinate points expected to follow.

5 *Use your notes in your study.* To retard forgetting, look over your notes soon after the class, concentrating on the important points. Fill in any gaps in your outline by recalling in your own words the missing material. Test your comprehension by formulating questions based on your notes. Use your notes in doing assignments and solving problems.

6 *Improve your note-taking skill.* Take a look at some other students' notes. Is your technique adequate? Compare your current notes with those you took several weeks ago. Are you improving? What is the greatest weakness in your current set of notes? How can they be improved? Along with reading and problem solving, note taking is a basic tool for engineers; be sure that your skill is developed to the point where you are obtaining maximum benefit from your lectures.

STUDYING THE TEXTBOOK

A textbook is especially written to serve as the basis for a course of study. Texts differ from popular books in that they tend to be concise and intellectually demanding; it is assumed that they will be studied carefully. Much of your time in college will be devoted to studying textbooks. To gain the maximum benefit from your hours of study, you should have the proper physical conditions and an efficient study procedure that takes advantage of what is known about effective learning.

Physical Conditions

Efficient study requires a 100 percent effort; therefore, all distractions, annoyances, and discomforts should be reduced to a minimum. One helpful rule is to let your friends know that your study place is reserved for study and that you do your relaxing somewhere else. Your study room should be quiet, at a comfortable temperature, and well-ventilated. Since most studying involves vision, lighting is important; the light should be bright enough for easy vision and located to avoid glare or shadows. If your lighting is good but you experience headaches or dizziness, see words as blurred on the page, or seem to tire quickly, have your eyes examined without delay.

Your chair and table should be selected to provide optimum conditions for work rather than rest; a straight chair and a table of the proper height should be used. The chair should face a wall rather than an open door or window, and the desk top should be free from distracting souvenirs from your last party. Tools such as pen, pencil, paper, calculator, and drawing instruments should be of good quality so that you are not handicapped in your attempt to do good work.

Procedure

When you sit down at that comfortable desk in that quiet room with your study material around you, what do you do? Here is where a tested technique can help you. To illustrate how to get the most out of an exploration of new territory in the world of knowledge, let us consider the comparable procedures for taking a trip. Your assignment is to travel to a distant land, study the conditions there, and report back on what you have seen.

Your first step is to get a map to preview where you are going. Then you plan the trip, listing what you wish to learn. Thus properly prepared you make the trip, keeping your eyes and ears open as you go. As you encounter an important feature, you identify it and record it on color film. Upon your return, you show the color slides, recalling the details surrounding each interesting feature or incident. Similar steps constitute an effective study strategy that can be described in terms of preview, prepare, read, recite, and review.

Preview   A preliminary overall picture of what you are going to study is like a survey of rough terrain from a mountaintop before you get down into the canyons—like a helicopter view of a forest before you begin a walk through the trees. The first time you open your textbook you should take time to get such a perspective. Look at the title and the author's name and affiliation. (Most students know only the color of their textbooks.) Read the preface to learn why the author wrote the book, and study the table of contents to see what the book is about. Read the introductory chapter if there is one.

Before each study session make a preliminary survey of the material. Check the table of contents to see what comes before and after the chapter you are studying. Read the introductory paragraph, scan the entire chapter, and then read the summarizing section if there is one. If you have a good overall picture of your study assignment, what you read will have added significance and relevance and will be easier to master.

Prepare   For highest efficiency in learning, your mind must be actively participating. Therefore, prepare your mind by formulating anticipatory questions: "What am I going to learn?" "What will I be able to do when I have finished?" "What message does the author have for me here?" If there are study questions at the end of the chapter or if your instructor made a specific assignment on this subject matter, look these over before beginning your detailed study. When you have properly prepared your mind, you are in optimum condition for study because then you are reading for a purpose.

Read   With an overall view and a few landmarks in mind and with a good idea of what you are seeking, you are ready to take the trip. Reading is an important skill that can be improved with training. As you read, be alert for important points indicated by headings or italicized words. Look up all unfamiliar words and be sure that you understand them. Read actively; look for answers to your questions or the questions posed by the author or your instructor. If the book is yours, make marginal notes and in this way make its contents yours as well. Debate with the author; raise questions and see if they are answered. If the author refers you to a table or an illustration, study it to discover whether it supports the author's contention. If in a derivation certain steps are omitted, fill them in. (Warning! Some authors misuse the word "obviously." If there is a neat and simple explanation, they will include it; but if the explanation is a little involved or hard to handle, he or she may write "obviously . . .") In all your reading have in mind the use for what you are learning, even if it is just preparation for a subsequent recitation.

Recite   Presumably, the material you are studying is to be recalled and used at some future time. Recitation is a form of immediate use that greatly retards forgetting. Studies show that material that is read and immediately recited is retained much better than similar material that is merely read. Other studies show that up to 80 percent of study time may be spent profitably on self-recitation. When you have finished reading a significant unit (several pages of literature, a topic in your chemistry book, or a single statement of a theorem in mathematics), put down your book and "re-cite." A theorem that is to be learned verbatim should be recited verbatim, but in general it is better to state the substance of your reading in your own words.

As you recite on technical material, convert mathematical symbols into words and equations into sentences. Answer the questions you formulated in your preparation or solve the type of problem that might appear on an examination. Applying new ideas to practical situations is an excellent form of recitation, because only ideas that have been mastered can be applied. One of the principal benefits of recitation is that lack of understanding of an idea or inadequate grasp of a concept is quickly revealed. If the meaning of a passage is not clear, phrase a specific question covering the difficulty; frequently, just stating the question is a big step toward the answer. If you cannot clear up the difficulty, take the specific question to your instructor.

Review   Most students review course work right before examinations, but there is more to review than that. Properly used, reviewing contributes greatly to mastery. A review is somewhat like a preview in that it gives you the overall perspective or the big picture with everything in place; this type of reviewing should be done whenever you have finished a unit of study—a chapter, a major section of course, or the entire course. A review may involve rereading the text to freshen up details or rereading your notes in preparation for an examination. During a review you should again recite the substance of the material studied, use it to answer questions, or apply it to typical problems.

For your greatest benefit, you should review frequently. As soon as you have finished your evening's work in a given course, take a few minutes to review. Some students set aside a weekly two-hour period in which they review the week's work in every course. Along with a further review at 5- or 6-week intervals for midterm examinations, these procedures result in the same type of overlearning that enables you to recall, as if by magic, the words to a song that was popular many years ago.

Here, then, is an effective scheme for study: preview, prepare, read, recite, and review. If your present study methods are haphazard, this systematic application of fundamental principles can work wonders for you, not only by improving your grades but also by providing an exhilarating sense of intellectual achievement. However, mere awareness of the existence of such a procedure means nothing; to be useful the system must be practiced faithfully and mastered thoroughly. Are you willing to put out the little extra effort now to develop the efficient habits that will pay off over a lifetime?

IMPROVING YOUR READING ABILITY

For the typical college student, 80 to 90 percent of study time is spent on reading, yet most students are not good readers. Good reading consists in obtaining a maximum amount of desired information from a printed page in a given amount of time. The two important factors are reading speed and reading technique.

Increasing Your Reading Speed

If you have been using all your available hours in just reading your assignments, you may be wondering how you can find time to do the recommended previewing, preparing, reciting, and reviewing. One obvious solution to your problem is to learn to read well; the time saved in reading will permit the other valuable study procedures.

But you say, "I already know how to read; I've been doing it for years." You know how to run, too, but that doesn't make you a runner. Reading is a learned skill and you are probably reading far below your capacity. With a real desire to improve, an understanding of a few basic principles, and some consistent practice, the chances are that you can double your reading rate and improve your comprehension at the same time. This has been the experience of many college reading clinics. If your school has such a clinic, arrange for a reading rate and comprehension test; if not, you may wish to study one of the references at the end of this chapter. As a starter here are a few specific suggestions.

Make a determined effort to read more rapidly. Force yourself to read at such a rate that you feel you are missing something; you probably aren't, because much of the printed page is not necessary for comprehension anyway. If you miss something go right on; subsequent phrases will probably fill in the gap. Rereading words is usually just a nervous habit and it is a great waste of time. Fifteen minutes a day of forced-draft reading of easy material will work wonders in improving your speed.

Read for ideas in phrases rather than word-for-word. We read by fixing our eyes on a point in a line, gathering information, and then flashing to another point. Poor readers have many fixations per line; they—read—like —this—by—look—ing—at—each—word. Good readers have few fixations per line; they look—at the center—of a phrase—and gather—a lot of information—at each flash. They read selectively, picking the meat out of a sentence without paying much attention to the little words that you leave out when you write a telegram message.

Eliminate any physical movements as you read. Don't form the words with your lips or in your throat; it takes twice as long to read something aloud as it should for silent reading and just forming the words slows you down. Hold your head still; let your eyes do the moving. The best readers seldom move their eyes clear out to the end of the line; for practice try reading a newspaper column by fixing your eyes only once in the center of each line.

Time yourself and keep a record of your reading rate. Select some easy fiction and time yourself, for, say, five pages. Strive for rapid reading and then later on time yourself on another five pages. Make a chart to show progress. Do not be discouraged by plateaus in your progress chart. Time yourself in reading an assignment in, say, history; then compete with yourself each week, recording the results. Bear in mind that average freshman students read familiar material at about 200 to 250 words per minute and that good readers do twice as well.

Improving Your Reading Technique

Increasing your reading speed is only half the secret of efficient reading; how you use your reading speed is just as important.

Read with a purpose. Ask yourself why you are reading a particular assignment. If you read to answer questions, your reading will be purposeful and therefore more productive. If you are looking for specific points of information, take advantage of the author's help. Use the headings set in distinctive type and the key words, which may be italicized. Spot the topic sentences, which usually occur at the beginning of a paragraph or, less frequently, at the end of a paragraph. Realize that most of the rest of the paragraph is just supporting evidence or clarifying detail.

Vary your rate of reading to suit the material. To find the meaning of a poem may require that you ponder each phrase or even read aloud with the proper tone of voice and rhythm. To locate a certain key fact, you should skim rapidly over the page. To benefit fully from a mathematical presentation, you may have to insert steps that are not there. To understand the plot of a piece of fiction, you can swing along at your maximum rate. In any reading task, adjust your reading rate to the difficulty of the material and to your particular requirements.

Keep building your vocabulary. In your reading, do you frequently encounter words that you recognize but only half comprehend, can't pronounce, or could never use even if you could pronounce them? Words you do not understand in a sentence are like chuckholes in an otherwise smooth highway: They slow you down.

It is axiomatic that good students have large working vocabularies. For this reason, most admission examinations include vocabulary tests. If you want to be a really good student as well as a good reader, get a dictionary and use it. When you encounter a new word, look it up, write down the word and its meaning, and use the word. In your engineering courses, mastery of technical terms is all-important; you can save time and increase your comprehension by learning the meaning of each new word as you meet it. For your general courses, you should be able to determine the meaning of many new words by analyzing them into prefix, root, and suffix; for example, *intractable* means "not (*in*) having the characteristic of (*able*) following (*track*)"; that is, unmanageable.

TAKING EXAMINATIONS

Before you can do a good job in anything, you must have clearly in mind what you are trying to accomplish. Elementary school teachers have been known to give tests as punishment for unruly classes, and many college students think of examinations as a form of torture that must be endured if they are to remain in school. But what are the underlying educational purposes of examinations?

Purpose of Examinations

To obtain an objective view of examinations, consider the case of part-time students who enroll in an evening course in computer programming at a community college. They pay for the privilege of attending lectures three nights a week, studying another three evenings a week, and taking examinations periodically. The programming can be used in preparing themselves in new areas required for advancement in their jobs. But what do they gain from the examinations?

In the first place, an examination is an instructional device; properly constructed, it indicates what is important in the course and illustrates how the facts, concepts, and principles are to be used. A good examination forces students to use new reasoning techniques and new approaches so that taking the examination is a learning experience. Second, an examination provides an appraisal of the student's mastery of the subject matter of the course and the ability to recognize, recall, or apply the material to new situations; it is important that students, and not just the instructor, have this evaluation of their progress. A third function of an examination is that it points the direction for further learning. Are the examinees ready to move ahead? Do they need more drill to develop speed and accuracy? Have they failed to master certain abstract concepts? Does the examination reveal unsuspected strengths and weaknesses suggesting a major shift in the direction of their study? In accordance with the principles of learning, the examination also provides an incentive to strive onward, so that by a certain time and under given circumstances a satisfying level of achievement will be reached and recognized. Lastly, and only incidentally, examination results provide a basis for comparing one student's degree of mastery with the established standards or with the mastery levels of other students, in other words, a grade.

Attitude toward Examinations

Are you willing to accept the foregoing statements as indicating the functions of the examinations you take? In what respects do you disagree? Insofar as they do apply, each of these points should be reflected in your attitude and approach toward examinations.

View the examination as another learning activity—competitive, but not an ordeal. Study your midterm examinations and other examinations from previous years to grasp the instructor's idea of the essential content of the course. While taking the examination, use the new methods and recently presented ideas instead of falling back on approaches you have learned in previous courses. Benefit from your hour of effort by learning during the examination.

Study your returned examination paper to see where you were well prepared and where you were weak. Compare your performance on several similar tests to see if you are improving in areas of weakness. If you are having difficulty in a course, analyze your test papers to see if you lack some basic ability, have an inadequate background, work too slowly, or make too many manipulative mistakes. Discuss your diagnosis with your instructor. Interpret your examination results in terms of suitability for your present vocational goal and discuss your findings with your faculty advisor.

Accept the examination as an incentive to learning, and plan your preparation accordingly. Organize your review to take advantage of distributed practice. As you study each topic, have in mind a purpose reflecting the type of examination you expect. Practice with old examination questions or extra review problems to discover where additional study is necessary. Adequate preparation will calm your preexamination nervousness, but accept the fact that an examination is a competitive situation and being keyed up a little is an aid in doing your best.

Consider the grade as only an indirect and inexact measure of your achievement and be neither unduly depressed nor excessively elevated by a single grade. A consistent pattern of grades is much more meaningful and should be taken into account as you make plans for the future. Tests and grades also serve to motivate you to study, particularly subjects that may be important but are not interesting to you.

Just what do grades measure and why are they important? A grade on an examination can only measure your performance in the activities required in the test. Your grade depends on your ability to interpret questions and follow instructions, your knowledge of the subject matter, your ability to apply that knowledge in a particular situation, and your skill in presenting that application for your instructor's appraisal. Course grades are important because they become part of your permanent record of academic accomplishment, and they are of interest to employers because they are accepted as measures of ability in engineering course work. It is recognized that there is a big difference between college courses and work on the job, but grades constitute a positive record of ability in one aspect of engineering. The pattern of grades is more important to an employer than the overall average. A student with low grades in advanced mathematics and science will have difficulty in obtaining a position in research. A student with D and F grades, even if they are balanced by A's and B's, gives the impression of a spotty performance.

Practical Suggestions for Taking Examinations

Since grades are important, it behooves you to take advantage of time-tested rules for getting better grades on examinations. Some of these suggestions are based on a careful analysis of what you are trying to accomplish in an examination, others are the result of psychological research, and still others come from inside information on how instructors think. Many good students have already discovered for themselves similar grade-building techniques. They will not turn a C student into an A student, but they may enable B students who have been getting C's on examinations to obtain the grades they deserve.

General   Convince yourself that if you are properly prepared you will do well on any examination, and prepare yourself accordingly.

Prepare in advance and allow some time for rest and relaxation so that you enter the examination room in your best mental condition. (An all-night cramming session before a 1-hour examination is as senseless as a mile training run just before a 100-meter dash.)

Develop a procedure for taking examinations so that you will be ready to start as soon as you receive the questions.

Look over the entire test and gauge the length and difficulty; allocate time for each part.

Always answer the easiest question or solve the easiest problem first, and leave the most difficult (for you) until last.

Read the instructions for each part carefully and be sure that you understand them.

Consider each question in the light of what you have read in the textbook or heard in the class; argue your personal opinions in the instructor's office, not in the examination book.

True-False   Read each statement through carefully, mark it quickly, and do not change your answers.

If there is a penalty for wrong answers, skip debatable statements to save time for other questions.

If there is no penalty for wrong answers, always guess on debatable statements.

Do not spend too much time on any one statement; come back to difficult questions later if you have time.

Do not change your answers; statistics show that you usually lose.

Multiple-choice   Follow directions carefully; you may be asked to choose the *most* correct item or the *least* correct item.

Read each statement, try to complete the statement properly, and then look for your answer in the list.

If you cannot find your answer, decide quickly on the proper item if you can and mark it.

If you are undecided, eliminate obviously improper items and try again.

If you are still undecided, mark the question in the margin and go on to the next statement; return if you have time.

Essay   If there is a choice among questions, make your decision quickly and get to work.

Make a brief outline of your answer before you start to write; put the outline in your examination book. This is very important; it will help you to stick to the subject in question.

If you can formulate a concise answer, put it down in compact form; don't pad it, unless you have reason to believe your instructor likes quantity.

Even if you feel you cannot answer the question, write something; frequently you do know enough to earn partial credit.

Learn what is meant by the key instructions: compare, contrast, define, explain, list, outline, prove, and trace. (See Assignment 6.)

Check over your paper to catch careless mistakes, unclear answers, or omitted questions.

Problem   Where appropriate, draw a clearly labeled sketch to help you visualize a word problem.

Where possible, draw a sketch to estimate an approximate answer for use as a check.

Clearly state any assumptions you are making in your solution.

Write clearly and show all your work so that the instructor can give you partial credit for method even if your result is incorrect.

Take advantage of dimensional checking and indicate the proper units for your result.

Save time to check your work for careless mistakes, unreasonable results, or omitted problems.

COMPLETING ASSIGNMENTS

Learning by doing is an excellent method of training. The best way to learn to solve mathematics problems is to solve problems. The best way to learn to write themes is to sit down and write. Going over and over the theorems of mathematics or the rules of composition will not bring the results that you can obtain by actual practice. One reason that engineering students master as much as they do is that the classroom presentation of principles and concepts is usually followed by homework assignments that provide valuable opportunities for learning by doing. Many of your assignments have additional value in that they provide practice in forms of communication that are particularly useful in engineering. In most assignments the ability to follow instructions is important; you can increase your future employability now by developing your skill in interpreting requests and following orders.

Endeavor to do *all* assignments *on time.* The ability to get things done completely and on time is particularly sought after by a project leader selecting members of a team, but good work habits are formed early and homework assignments provide excellent practice. Do all the assigned problems even though the homework score does not count much toward the final grade; take full advantage of this learning opportunity which is the best possible preparation for the examinations that do count. Don't get behind; actually it takes less time to do an assignment on time because the material is fresh in your mind. Try to stay within your study schedule, working faster if necessary; increased speed will pay off when you are solving similar problems on an examination.

In addition to handing in the results of your work, you should strive for an effective form of presentation. If your instructor has specified the form for homework assignments, follow these instructions carefully. If not, the following suggestions based on good engineering practice may be helpful:

Use good quality paper. The type that is quadruled on the back makes it easy to do a professional job on problems because the lines serve as lettering guides and facilitate sketching diagrams and drawing graphs. For typing, use an erasable bond.

Label your paper. Include your name, course number, section number or instructor's name, and the date. Put this information on the outside if the paper is to be folded.

Don't crowd your work. Paper is relatively inexpensive, and taking enough space for a neat layout will pay off in appearance. Start each problem on a new sheet unless it is obvious that two short problems will fit on a single page. Use only one side of each sheet.

Use a medium pencil. Develop the habit of careful, logical presentation as you carry out your solution. Show all your work to provide a complete permanent record, but you may wish to do intermediate calculations in the left margin. Don't scribble on scratch paper.

State the problem in sufficient detail. This will enable you to review your solution sheet effectively without reference to a separate assignment sheet.

Use sketches, diagrams, and graphs in your statement of the problem, in your solution, and in presenting your results. Graphical solutions are particularly valuable as checks.

Practice lettering instead of using cursive writing. Many engineers develop a "printscript," which becomes as fast as longhand but has the neat professional appearance of lettering. In printscript, letters are formed individually, but by continuous strokes with some compromise in precision to gain speed.

State the principle being applied or write the general equation before substituting specific numbers. Make your solution read coherently by indicating briefly the reasoning behind each step.

Use the equals sign correctly. Be sure that both quantities are truly equal. If you wish to take the result of one calculation and perform a subsequent operation, start a new line.

Designate your results clearly by underlining or boxing. Significant intermediate answers can also be indicated. Be sure to include the proper units.

Every assignment is a report to the instructor on your progress—not only your progress in mastering new material but also your progress in developing skill in communication. A professional-appearing paper says to the reader, "Here is the work of a student who can solve problems, who appreciates the importance of a good presentation, and who is willing to make the extra effort to do a superior job." When you place your signature on a homework paper is it with a feeling of pride in an assignment well done?

SUMMARY

The four or more years that you will spend in college are far too short a time to master all that you need to learn; it is essential that you use the time available now to the best possible advantage. Also, upon graduation you will realize that the need for studying continues; in this age of spectacular scientific breakthroughs and continuous technological advances, the engineer who stops learning is quickly passed by. For at least two reasons then, you need to develop your study skills to the utmost.

Good study habits are based on a knowledge of the principles of learning, application of these principles in the formulation of effective study procedures, and the practice of these procedures in the development of the basic study skills.

You will learn better and faster if you are properly motivated, if you concentrate all your attention, if you comprehend what you are studying, if you study actively and eagerly, if you distribute your learning over a period of time, and if you provide a means for evaluating your achievement.

If you have difficulty organizing your time, a time budget will help by eliminating indecision about what to do next, by turning odd hours into solid achievement, by encouraging efficiency through scheduling, and by putting the time you have available into the things you wish to do.

A good set of lecture notes is an invaluable study guide, and the process of taking notes is itself an aid in learning. Develop a good note-taking technique and perfect it by regular practice.

An effective study procedure consists of *previewing* to gain the proper perspective of what you are going to study, *preparing* your mind for active participation by formulating questions, *reading* with a purpose and with your mind alert for important points, *reciting* to test your mastery and to retard forgetting, and *reviewing* frequently to increase your learning efficiency.

You can greatly increase your reading speed by forcing yourself to read faster, reading for ideas in phrases, eliminating physical movements, and keeping a chart of your progress. You can improve your reading technique by reading with a purpose, varying your rate to suit the material, and continuing to build your vocabulary.

Examinations are here to stay; try to obtain the maximum benefit from them, and learn how to do your best in them. Don't be too concerned with individual grades, but accept the fact that your pattern of grades is a valid measure of certain abilities. Remember that as a young engineer you will be examined and graded continually.

Consider each assignment as an opportunity to learn by doing. Develop the habit of doing assignments completely and on time. Take pride in your work and enjoy the satisfaction of achievement. When the work of learning is well organized, the joy of learning becomes apparent.

You can forget every rule in the book if you will remember one thing: Good work habits result in achievement; achievement brings satisfaction; satisfaction generates motivation; and motivation increases achievement.

REFERENCES

1 Armstrong, W. H., *Study Tips: How to Study Effectively and Get Better Grades,* Barron, New York, 1975.
2 Brown, G. E., *A Student's Guide to Academic Survival,* Harper & Row, New York, 1976.
3 Coffman, Sara, *How to Survive at College,* TIS Enterprise, 1981.
4 Erasmus, John, *How to Pass Examinations,* 3d ed., Routledge, London, 1977.
5 Gross, Ronald, *The Lifelong Learner,* Simon & Schuster, New York, 1977.
6 Vogel, J. J., *Study and Like It,* Exposition Press, Hicksville, N.Y., 1977.
7 Yarington, D. J., *Surviving in College,* Bobbs-Merrill, Indianapolis, 1977.

ASSIGNMENTS

1 Prepare a time budget for your current program by following the procedure outlined in this chapter. At the top of the sheet, present an analysis of your courses in terms of their time requirements and add in reasonable allowance for work, recreation, meals, transportation, etc. On the lower portion of the sheet, show a tentative weekly time schedule developed in accord with principles of effective learning.

2 Conduct the following experiment. During a 1-week period, allow yourself just 24 hours of outside-of-class study, but plan your work to use this time efficiently. Describe your plan and report on how well it worked in comparison to your previous approach.

3 (a) Cite specific examples showing how you are now taking advantage of each of the six factors in effective learning. (b) Cite specific examples of how you could be taking better advantage of the six factors.

4 Take a typical page from your lecture notebook and analyze it in terms of the comments made in this chapter. Point out major weaknesses and major virtues. List specific ways in which your notes could be made more effective. Submit your notebook page and your analysis for review by your instructor.

5 Conduct a reading improvement project over a 3-week period. Select material from a history or literature course, for example, that will not change much in character over the period. Determine your base reading rate by averaging the results of three separate tests on five-page samples. Conscientiously apply the suggestions for increasing your reading speed and retest yourself three or four times each week under similar conditions. Record the results of your experiment on an appropriate graph.

6 Look up in a dictionary the following key words which frequently occur in examination questions: compare, contrast, define, describe, diagram, discuss, enumerate, explain, illustrate, list, outline, prove, trace. On a page of your notebook paper, record the definition which is most applicable to an essay question.

7 Devise a study plan of action for the course that is giving you the most trouble. Translate the general principles given in this chapter into specific plans for improving your performance in this most difficult course. Include suggestions to yourself regarding lectures, studying, assignments, and examinations.

11

WRITTEN AND ORAL REPORTS

For many years you have been working at developing your ability in self-expression. As a small child you quickly learned to express your ideas and wishes orally. In your first years of school, you learned to associate ideas with abstract symbols—words. Then, wonder of wonders, you learned to write your own mystic symbols that others could translate. Later on you began to analyze sentences; you learned the parts of speech—the nouns, verbs, adjectives, adverbs, etc.—and the function of each. Understanding the functions of the component parts, you were able to do a better job of synthesis—putting the parts together. Your subsequent training in English probably emphasized the clear expression of ideas with continued practice to help eliminate mistakes in the mechanics of writing—spelling, punctuation, grammatical construction, and word usage.

In spite of all the training students have had before entering college and the instruction and practice received in college, a common complaint of employers is that their new employees cannot write satisfactorily; they cannot compose simple business letters, for example, without glaring errors. Perhaps one reason for this is the students' lack of motivation; they have not been made to realize the importance of expression skills in their own careers.

The purpose of this chapter[1] is twofold. First, it will call your attention very early in your college training to the importance of verbal expression in engineering and indicate its relevance to your future progress. Second, some preliminary training will be given in the form of expression so commonly used in engineering—the writing of reports. The emphasis in this discussion will be on the type of planning called design, which is found in all engineering work.

THE IMPORTANCE OF SELF-EXPRESSION IN ENGINEERING

Engineering is essentially a mental activity, and the results of engineering work are usually in the form of ideas. For example, the result of a year's research work may be a new concept of the heat transfer mechanism in a molten metal. An industrial engineer may strive to create a new and satisfactory approach to incentive pay. (You should be able to cite similar examples from each of the other engineering functions.)

To be useful, these ideas must be transmitted to others for their consideration or action. The engineer may be called upon to express his ideas while:

- Reporting to a superior and to other departments
- Discussing instructions received from superiors
- Giving orders to subordinates
- Arguing for a proposed plan
- Writing contracts and specifications
- Writing operation and service manuals

Although all these examples involve the communication of intelligence *verbally,* by means of the spoken and written word, engineers are unique among professionals in that they must also be able to express themselves *graphically,* in terms of charts or drawings, and *mathematically,* in terms of numbers and symbols.

The importance of skill in communication cannot be emphasized too strongly. From today on, through four years or more of engineering training and throughout your entire engineering career, you are going to translate the output of your mind into usable form. For some of you the process of expression will be a bottleneck—restricting and distorting your mental efforts. For the more fortunate, the process will be a filter for rectifying, clarifying, and enhancing your ideas.

---

[1]Much of this chapter is based on "Writing the Technical Report," by J. Raleigh Nelson, McGraw-Hill, New York, 1952. Although this excellent book is out of print, a copy may be in your college library.

You should realize that every assignment and every examination is a report to your instructors or a demonstration of your progress along the lines of their instruction. In the same way, each oral and written report to your superior will be a manifestation of your thinking ability. In many cases your promotion or assignment to a position of increased responsibility will be decided by an executive whose knowledge of your ability is gained primarily from your written reports.

DEFINITION OF A REPORT

The word *report* means to "carry back," that is, to bring back information about something seen or investigated. So the engineering report, oral or written, is a means for making available to the listener or reader information in which he is interested and, frequently, information which he has requested. Implied in this statement are three characteristics of a report: It is a form of what is called expository writing; it is written for a definite purpose; and it must be in a form suitable to the reader.

A Form of Exposition

In contrast to other forms of writing such as description, narration, or argumentation, exposition is the setting forth of ideas or facts in the most revealing manner possible. Just as the exhibits at a world exposition are carefully and systematically organized to present information about products, so the portions of a report should be planned to make the transfer of ideas easy. In the whodunit, the writer seeks to mystify and confuse the reader until, in the final courtroom scene, all is disclosed. In the report, however, the writer strives at all times for clarity.

To achieve clarity, the material in the report must be organized according to a definite plan and accompanied by the necessary explanation and interpretation. The organization must be consistent with the underlying purpose of the report. In addition to presenting the bare facts or ideas, the report must anticipate questions, describe equipment or methods used, explain technical aspects or complex relationships, and interpret observations and results.

Designed to Accomplish a Definite Purpose

One distinguishing characteristic of the report is that information is carried back for a reason; usually, the reporter is sent out in the first place because of a need for certain information. That information may be needed in making a decision—for example, a report on the desirability of changing over from direct-current power to alternating-current power. Or the information may contribute to a critical evaluation of an operation—possibly a report on the annual cost of operation of a metropolitan water system. Or the information may be in the form of a request for action on a proposal—why the manual filter control should be relaced by an automatic unit.

Engineering design consists of starting with the specified end result and working backward. The engineer designing a portable radio receiver, for example, starts with a set of specifications in terms of desired volume and tone output and limitations on weight and cost. To handle the desired signal output and frequency response requires a certain type of speaker which must be fed by a suitable audioamplifier. This, in turn, dictates the required power supply, the detector circuit, the radio-frequency amplifier, and so on, clear back to the antenna and the input signal. In exactly the same way, the designer of a report must start with the desired end result and select the components and methods of presentation that best will accomplish that purpose.

Designed to Meet the Requirements of the Reader

Throughout the planning of the report the characteristics of the reader must be kept in mind. A newspaper story or a magazine article must be designed for the general public with eye-catching headlines and suggestive pictures. In contrast, the report is written for a particular reader or a small group of readers whose requirements are known.

The characteristics of the reader will determine the length, technical level, and plan of the report. Busy executives demand from their technical assistants brief, concise reports with a minimum of detail, whereas the board of supervisors requires of its consultants on mass transportation a complete discussion of technical, economic, social, and legal aspects which may take thousands of pages. A report to the chief development engineer can be on a scientific level; the plant operator will be more interested in technical aspects and can understand the technical terminology which would be undesirable in a report to a city council. The plan also will be dictated by the characteristics of the reader and the desired result, with the particular approach and sequence of topics carefully selected.

PROCEDURE IN WRITING A REPORT

So far, we have discussed why reports are important in engineering and what is to be accomplished in a report, but just how do you go about writing a good report? Fortunately, it is not as difficult as you might think. The procedure recommended here consists of four distinct and separate steps: studying, designing, writing, and checking.

Study the Material

Usually a report is written after some engineering work and is based on information gathered by observation, investigation, measurement, or experimentation. The first step consists of assembling, sorting, and appraising the available information and data. There may be some calculations, graphs, pictures, tabulated data, observations, expert opinions, or other reports on the same subject. What do the data indicate? What do you have to tell? Is your information adequate, or do you need additional data? Until the answers to these questions are clearly in your mind, no attempt should be made to begin the second step.

Design the Report

Now comes the planning. What is the purpose to be accomplished in the report? What are the limitations on your discussion, or what is the precise subject? What is your viewpoint on this matter? What are the characteristics of the reader, and how will these affect the handling of the report? The answers to these questions indicate the strategy to be followed in presenting the available information in order to accomplish the desired objective. There is a close analogy to military strategy in which the available personnel and weapons must be distributed and their advances timed to accomplish a given objective with a minimum loss.

This design stage is probably the most important of all, and it merits considerable time and effort because it is the quality of the design that determines the quality of the report. It is important that the design be complete before you do any writing. Nelson suggests two conclusive tests of progress, the *thesis sentence* and the *outline plan.*

The thesis sentence is a brief statement of just what you have to say; it is the message you have to report. For example, the thesis sentence for this chapter might be: The oral and written reports that are so important in engineering should be designed to accomplish a definite purpose and to meet the reader's requirements.

Until you can write out a single sentence that clearly indicates the precise subject to be presented and your viewpoint on that subject, you are not ready to write. The subject is limited to oral and written engineering reports, eliminating, for example, poems. The viewpoint is "reports . . . should be designed". There is also a hint of the plan of the chapter—first a discussion of why reports are important in engineering and then some training in report design.

The outline plan indicates the structural pattern of the report and is based on an analysis of the material available in terms of the purpose to be accomplished. The outline indicates the important topics, sequence of topics, and the relationship between major and minor topics. One good approach for you as a beginner is to enter on individual 3- by 5-inch cards the bits of information and data resulting from the investigation. Then, spread the cards out on your desk; with your thesis sentence always in front of you, sort and shift them until you obtain an effective organization of material. After you have arranged the cards according to the most logical and strategic plan, you can transfer the topics to a single sheet of paper. The outline plan for the first half of this chapter was as follows:

I Introduction
    A Subject: written and oral engineering reports.
    B Purpose: to provide motivation for studying English and some preliminary training in report writing.
    C Plan: as indicated in the outline.
II The importance of self-expression in engineering
    A Engineering work usually results in ideas.
    B To be useful, these ideas must be communicated to others.
    C The engineer must be able to express his ideas in three ways: verbally, graphically, and mathematically.
III Definition of a report
    A A report is a form of exposition.
    B It is designed to accomplish a definite purpose.
    C It is designed to meet the requirements of the reader.
IV Procedure in writing a report
    A Study the material.
    B Design the report.

Actually the outline also included third-order headings, subheadings 1, 2, and 3 under second-order headings A, B, and C. Check the first part of the chapter against this outline and then, to see if you understand outlining, write out the outline for the remainder of this chapter.

Write the Rough Draft

Don't begin to write until the design is complete—until you have a thesis sentence and an effective structural plan. Once you start to write, emphasize production—getting your ideas down on paper—and leave criticism of your work until a later time. If you attempt to perfect each sentence as you go along, you will interrupt the smooth flow of ideas which is so important in a good report.

Another way of achieving a smooth flow of ideas is to write a relatively large section at one sitting with no interruptions; this will tend to produce coherence—the ideas will hang together. Some of you will be able to write rapidly *and* well, but this usually takes years of practice.

Check the Report

The rough draft must be checked against the design, and it must be compared with high standards of composition. To determine if the design has been carried out, you should ask yourself such questions as: Does it accomplish the stated purpose? Does it meet the particular needs of the reader? Does it stay within the limitations of the subject?

The first draft of a beginner will probably be pretty rough, but if it is unified and structurally sound and if the ideas flow smoothly, the final polish is not difficult. Is it free from distracting mistakes in grammar, word usage, punctuation, and spelling? Are the sentences strong, clear, and effective? Are the paragraphs consistent and coherent? Have you kept your standards high? If so, you and the reader (possibly your employer) will be well pleased with the result.

GENERAL ORGANIZATION OF A REPORT

In setting up an exhibit at an exposition, every effort is made by the exhibitor to make it easy for the viewer to gain an understanding of the display. So in a report, the test of every part is asking the question: Does this help the reader? There are certain standard methods of helping the reader with the introduction, the body, and the terminal section.

Introduction

The introduction is a preliminary conference between the reader and the writer. In a sense you say to the reader, "Before we get into the main part of the report, I am going to indicate just what the report is about, explain what my attitude is toward the subject, and point out how I am going to present the material." In general, the introduction should include the subject, the purpose or viewpoint, and the plan. If you define the limits of the discussion, the reader will not be disappointed or critical if certain aspects of interest to him are not discussed. If, before he gets into the detailed discussion, the reader has a picture of the plan to be followed, he will not get lost. If he has a clear idea of what you want him to get out of the report, the reader will be looking for your points and give them his full attention.

Body of the Report

Working from the previously designed structural plan, each topic heading from the outline is translated into a section of the report. Paragraphs can be introduced by topic sentences, which are then amplified and explained in the remainder of the paragraphs. In a sense, the writing consists of hanging meat on the skeleton.

To help the reader, the writer should make clear the change in thought represented by a new section and, at the same time, he should indicate how the new material is related to what has gone before. The phrases and sections which serve this purpose make up the transition material. The paragraph on page 257, beginning "So far..." and ending "... and checking" is a transition section. How does it indicate the beginning of a new topic? How does it relate the new material to what has gone before?

Another way to help the reader is by using section headings. These are used in textbooks and magazine articles, as well as in reports, because the reader at a glance can know the topic under consideration. Frequently the outline plan can be converted into a schedule of section headings, although rewording is usually desirable.

Terminal Section

The longer the report, the more attention should be given to keeping the reader informed as to what the next topic is to be and how it is related to what has gone before and what is coming later. However, it is not possible for the reader to keep all the points in mind as he or she reads through the report. The terminal section serves as a final conference with the reader in which the writer can bring the discussion to a proper close consistent with the original purpose indicated in the introduction.

The terminal section serves to bring the essential material into focus in a single section and to make sure that the reader gets the essential idea. The human mind is limited in the number of ideas that it can grasp at one time; there are indications that the maximum number is around five to seven. Certainly no one can keep clearly in mind thirty different aspects of a problem. There is also a limit to the quantity of material you can retain in mental focus—probably less than a page. As you read through a multipage report or a chapter in a textbook, early points fade into the background as additional ideas take their places. The terminal section provides an opportunity to create the most desirable final impression.

The terminal section takes different forms depending on the purpose of the report. A *summary* is a restatement of the important points with the mass of detail omitted; it usually parallels the presentation in the report. A *recommendation* consists of suggestions for action based on the discussion and may include a brief statement of the reason for each recommendation. In a *conclusion,* the writer draws conclusions from the evidence submitted in the body of the report; no new ideas should be introduced in the terminal section.

What is the most appropriate type of terminal section for this chapter? Referring to the introduction, you will see that the purpose was to provide motivation for studying verbal expression and to give some preliminary training

EXAMPLE A

The business letter is a special type of report following a conventional form. Note in this application for employment the three paragraphs representing introduction, body, and terminal section. Even in this brief introduction, the subject, purpose, and plan are indicated.

                                                    2912 Newbury St.
                                                    Tacoma, Washington
                                                    April 13, 19—

Mr. Eric Jason, Personnel Manager
Judson Electronics Company
7265 Willamette Blvd.
Seattle, Washington 98199

Dear Mr. Jason:

    I am a sophomore in electrical engineering at Rainier State College, and I am interested in obtaining summer employment with your company. Perhaps my background of experience and training would make me a useful temporary employee.

    Four years ago I became interested in radio, joined my high school radio club, and obtained my amateur license. I built my own equipment and have been active as a "ham" ever since. I repaired radio and television sets for my family and friends to finance my hobby, but my principal interest is the engineering aspects of electronics—specifically development work. So far I have completed two years of mathematics and one year each of chemistry and physics, all with grades of B or better. The schoolwork is very interesting, especially the electrical part of physics, but I am eager to obtain more practical experience.

    Judco products are well-known among radio amateurs, and employment with your company would provide an excellent opportunity to learn what is behind a quality product. I realize that my background would fit me for only a routine test or assembly job, but I believe that you would be pleased with my work in such a position. I will be in Seattle on Thursday, April 30, and would appreciate an interview at a time convenient for you.

                                                    Sincerely yours,

                                                    Kendra N. Tristan

---

EXAMPLE B

The memorandum is a modified form of a short report that is used frequently in business and industry. It resembles the letter, except that the salutation and complimentary close are omitted and a standardized heading is used. The elements of a good report should be retained although in abbreviated form. The "subject" is stated clearly and completely, and the "viewpoint" or "purpose" indicated; the "plan" may not be necessary in a short "memo". The necessity for careful organization of a short report was emphasized by the writer who said, "Please excuse this long letter; I did not have time to write a short one."

INDIANA AIR TRANSPORT, INC.
Interdepartmental Memorandum

DATE: October 25, 1981

TO: Thomas Gregory, Engineering Department
FROM: Rachel Gabriel, Safety Engineer
SUBJECT: Hangar Fire October 21: Cause and Recommendation

We have completed our study of the fire which occurred in aircraft 670 last week during a run-up check of the power generating system. It is our finding that intense arcing of a loose electrical ground connection set fire to spilled hydraulic fluid, and a redesign of the ground connection is indicated.

When the connection to ground was made, a slightly longer bolt was substituted for the correct part. Because of insufficient clearance, the longer bolt bottomed on the airframe member to which the ground strip was attached so that the torque wrench indicated a tight connection when in fact the electrical connection was poor. The connection was loosened further by vibration and the resulting arc ignited fluid spilled when the hydraulic lines were opened prior to the run-up check.

There are on record many similar cases of fires resulting from poor electrical connections, and in designing electrical equipment, special care must be taken to avoid the possibility of loose connections developing. Based on our recent experience, we recommend that the ground strip for this model aircraft be redesigned to provide studs instead of the threaded bosses now used.

---

in report writing. In line with that purpose, probably a brief summary would be the most effective way to close this section.

ORAL REPORTS

We are in the midst of an information explosion. It has been said that the average person today absorbs more than 20,000 words of information every day. A major problem is how to understand, organize, and communicate this information.

Most individuals, even those with a great deal of experience in public speaking, have anxiety attacks when called upon to give an oral presentation before a group. Knowing your material well and organizing it in much the same manner as you would a written report, as described above, will increase your confidence and contribute to a more relaxed presentation. Many organizations and educational institutions offer short courses and seminars for the purpose of helping individuals increase their oral communication abilities. Toastmasters International is an organization with branches in most communities whose goal is to help members, frequently college graduates, improve their public speaking skills.

Oral reports are generally for the purpose of providing information. Your ability to make an effective presentation of this nature can make a significant difference in your career. When presenting such information, it is important to keep your point of view in mind. Are you taking the role of advocate or are you being strictly objective? If you are dealing with a subject on which you have strong opinions, tell the audience of your bias at the beginning, so they can judge your conclusions with greater understanding. Some of the following points may be useful to you in preparing your material for an oral presentation:

1 Be relevant to the audience's experience and desires. It is up to you to discover the relationship between your topic and what the audience wants to know.

2 Organize your material into small, clear segments. You don't want to choke your audience with too much information. Break the material down into separate components, and present them in logical sequence. Let the audience know what your organization system is, so they can follow you (e.g., "There are three parts to this process. First . . .").

3 Present the information in an interesting manner. To keep the audience with you, inject a little humor, drama, or startling fact into your presentation. Try walking around a bit as you talk, to keep your listeners from developing a glassy-eyed stare.

4 Involve your audience. Throw questions out to them or have them write things down. When they participate, they learn.

5 Use as many channels as possible. Your words are only one way to get the information across. Involve senses other than hearing, especially the visual sense. Use chalkboards, overheads, slides, etc.

6 Associate the unfamiliar with the familiar. When people hear new information, they listen for familiar themes that will allow them to place the new information within their previous experiences.

7 Repeat important points, or restate them in different words. Allow yourself enough time to make your points as often as necessary to ensure understanding. You have to repeat and reinforce. Maintain constant eye contact to judge whether or not you are being understood.

Organizing your Presentation

How you organize your presentation will depend on what kind of information you are trying to communicate. However, one of four general approaches will probably meet your needs:

1 *Time sequence.* If you are describing a process, you will probably want to explain the beginning, intermediate, and final stages.

2 *Function.* Information on a product or service is often best given in terms of what each part does. A speech about your electronic calculator would include the various calculations or operations that can be performed and how they are accomplished.

A speech about the development of an engineering design for a new solar energy source might begin with the identification of a need; it might proceed by relating basic and applied research, developing and testing a preliminary model, preliminary and final design stages, production planning, quality control and inspection; it might end with a marketing and sales plan.

3 *Structure.* An organization would be best explained by how its various parts operate and how they interrelate.

4 *Deduction from principles.* If you want to explain an idea or concept, you might begin with the general principles, such as the relation between velocity and air pressure, and then show how these principles can be applied—to an airplane or a carburetor, for example.

Within these approaches, you will naturally want to develop an introduction, body, and terminal section. As was mentioned above, it is especially important for an informative presentation to be clearly organized, with the audience aware of its organizational structure. The body will normally consist of support material for the points you are making. Don't force the audience to merely take your word for the information you present. Support your points with facts, quotations, and the opinions of experts. This will help reinforce the audience's learning as well as convince them you know what you are talking about.

SUMMARY

The results of most engineering work are in the form of ideas that become useful only when they are communicated to others. You may find that your engineering ability will be measured in terms of your ability in self-expression, and your entire career may depend on your skill in making oral and written reports. College is the best place to develop that skill.

A report is a form of exposition that has been designed to accomplish a definite purpose and to meet the requirements of a particular reader. The preparation of a report can be accomplished in four separate steps. First, the material should be studied until you are sure of what you have to tell and that your data are adequate. Next, the report must be designed to accomplish the desired objective; working from a thesis sentence, you should develop a strong structural plan. The third step is writing, which should not be started until the design is complete. Once you start to write, emphasize production and leave criticism of your work until later. Finally, check the rough draft to see that it does what it was supposed to do and it meets the high standards of composition.

Throughout the entire report—the introduction, the body, and the terminal section—make every effort to keep the reader informed on what you are doing and why, so that when he or she finishes the last sentence, the full benefit of your work and your thinking has been obtained. In all your writing, keep your standards high; you will never exceed them.

An oral report is similar to a written one in that it is also a form of exposition that has been designed to accomplish a definite purpose and meet the requirements of a particular audience. Organization and knowledge of the material are equally important. The difference is that you are going to be judged not only on the basis of the content of the material presented but also on the basis of your personal delivery—how effective you are in verbally communicating your knowledge and information to your audience. Take advantage of college or community opportunities where you will receive friendly, critical advice as you gain speaking experience. The old adage "practice makes perfect" may not always apply, but certainly the more presentations you give, the more relaxed and effective you will become.

REFERENCES

1 Baker, S., *The Complete Stylist Handbook,* 2d ed., Harper & Row, New York, 1980.
2 Strunk, W., and E. B. White, *Elements of Style,* 3rd ed., Macmillan, New York, 1978.
3 Daiker, D. et. al., *The Writer's Option,* Harper & Row, New York, 1981.
4 Houp, K. W., and T. E. Pearsall, *Reporting Technical Information,* 4th ed., Collier-Macmillan, London, 1980.

5 Putnam, L. L., and R. E. Crable, *Principles of Human Communication,* Kendall Hunt, Dubuque, Iowa, 1980.

6 Monroe, A. H., et. al., *Principles and Types of Speech Communication,* 8th ed., Scott, Foresman, Glenview, Ill., 1978.

7 Simons, H. W., *Persuasion: Understanding, Practice, and Analysis,* 1st ed., Addison-Wesley, Reading, Mass., 1976.

8 Borman, E. G., *Discussion and Group Methods,* 2nd ed., Harper & Row, New York, 1975.

9 Swanson, Delia, *The Nature of Human Communication,* Science Research Associates, Palo Alto, Calif., 1976.

ASSIGNMENTS

1 Considering Chapter 5, "Branches of Engineering," as a report, what paragraphs constitute the introduction? What words or phrases indicate the subject of the report? The viewpoint? The plan?

2 Considering Chapter 8, "Statistical Picture of the Engineering Profession," as a report, what paragraphs constitute the introduction? What words or phrases indicate the subject of the report? The purpose? The plan?

3 Considering Chapter 7, "Creativity and Decision Making," as a report, write a thesis sentence and prepare an outline plan. Suggest another way in which this chapter could be organized to achieve the same purpose.

4 Study an engineering report or a magazine article based directly on a report. Cite the title, author, and source and answer the following questions:

a In what respects is the report an example of "exposition"?

b What is the definite "purpose" for which the report was designed?

c What are the characteristics of the reader for whom the report was written?

d Prepare a 3-minute summary to be given in an engineering class.

5 Study the technical literature to obtain a good understanding of a new construction method, production process, or maintenance procedure. Assume that you are the appropriate engineer and prepare a two-page memorandum recommending to your boss the adoption of a new idea.

6 Write your instructor a brief report in business-letter form describing your background, explaining how you came to study engineering, and indicating the type of work you hope to do in the future. Follow all the rules for report writing.

7 Assume that you are an application engineer recommending one of your company's products to a customer who is considering an entirely different product. Write a one-page report designed to convince your customer (who is a nontechnical person) of the advantages of your product. Submit a thesis sentence and an outline plan along with your report.

8 Assume that you are an engineer who has just returned from an exposition at which you saw displayed a new device that might be applied in your plant. Write a one-page memorandum to your boss describing the operation of the device, pointing out how it could be applied, and making a specific recommendation. Submit a thesis sentence and an outline plan along with your report.

9 Your boss has asked you to prepare a 5-minute oral summary to be given at your weekly staff meeting based on the device described in Assignment 8. Prepare an outline and give the presentation to a group of engineering students and/or faculty.

10 Prepare an application for summer employment in business-letter form, following all the rules for report writing.

11 You have been asked to report at the next student chapter meeting of an engineering society on why engineering seniors should take the Engineer-In-Training exam. Prepare an outline for your talk; then contact a professor and engineering society officer and ask permission to present the talk.

Some illustrative problems
Steps in the engineering method
Modeling and design
Application of the method

12

THE ENGINEERING METHOD OF PROBLEM SOLUTION

While on a vacation trip in the mountains you stop to look at the wreckage of a truck whose brakes failed on a long downhill stretch of highway. As you view the pile of twisted steel you imagine the scene at the time of the accident: a loaded truck careening down the grade, horn blowing, other drivers frantically trying to get out of the way. If you were a highway engineer, what could you do to reduce the chance of such an accident?

One activity common to all engineering work is problem solution. The problems may involve quantitative or qualitative factors; they may be physical or personal; they may require abstract mathematical and scientific concepts or merely common sense. The problem in the preceding paragraph is typical of some of the problems faced by engineers in their daily work. The solution requires the same approach as many more difficult problems. The ability to attack and solve problems is one of the most valuable technical skills, and much of engineering training is directed toward developing this ability.

The discussion of the engineering branches emphasized the results of engineering efforts. This chapter will attempt to explain how engineers go about obtaining the results. (Other professionals—doctors, lawyers, ministers, artists—employ entirely different approaches.) The assignments at the end of the chapter will provide an opportunity for you to try your hand at solving engineering problems. In terms of Table 12-1, these problems are isolated but not completely defined, involve few variables, and employ well-known approaches.

For the next 4 years you will be gaining a background of factual information, a knowledge of engineering principles, and skill in applying the principles to more complex problems. After graduation you will begin to gather the experience that will develop the judgment needed for solving real problems. The purpose of this chapter is to describe the general engineering method that can be applied to problems in many fields. It is believed that with this background you will be in a better position to benefit from your technical studies.

Let us consider some very elementary problems and try to discover the basic steps in their solution.

TABLE 12-1
LEVELS OF ENGINEERING PROBLEMS

Level	Degree of definition	Variables	Approach
I	Isolated and defined	Few	Known
II	Isolated and defined	Few Several	New Known
III	Isolated but not defined	Few Several	New Known
IVa	Isolated but not defined	Several Many	New Known
IVb	Neither isolated nor defined	Few Several	New Known
V	Neither isolated nor defined	Several Many	New Known
VI	Neither isolated nor defined	Many	New

In deriving an equitable basis for determining relative salaries, the director of executive development of a major corporation decided that the best measure of ability is the complexity of assignments that the engineer has demonstrated his or her competence to handle. In classifying problems by level, the director took into account the extent to which a problem has been located and defined, the number of variables or elements involved, and whether a new or unusual approach is required.

Source: Ralph W. Peters, Jr. "A New Approach to Engineer Salaries." *The American Engineer*, August 1967.

Problem A

What is the highest speed reached by a Boeing 747 on a regular flight between San Francisco and Los Angeles?

Well (you might say to yourself), there are at least three approaches to this problem: (1) I could call up the Boeing Company or the airline and ask; (2) I could take the trip and make my own observations by using a stopwatch and the road pattern in open areas where the roads border mile-square sections; (3) I could calculate an approximate speed from distance and time. The most direct method would be to ask someone who knows; upon calling the local travel agent, however, I find that she only knows that "it's a 45-minute flight." The time and expense of the second approach isn't justified under the circumstances. The third method is simple, and the approximate answer would be satisfactory.

To calculate the speed I need distance and time. I know it's 404 miles by highway from San Francisco to Los Angeles. On a 45-minute schedule, I visualize the airplane taxiing to the runway, getting permission to take off, accelerating down the runway, climbing to cruising altitude over the Pacific, flying at maximum speed down past the Tehachapi Mountains, slowing down to enter the flight pattern, landing, and taxiing to the gate. If I assume that the distance by air is 10 percent less than the distance by highway, that takeoff and landing each consume about 5 minutes, and that the remainder of the flight is made at constant speed, the problem can be solved.

The net time available for cruising is then $45 - 5 - 5 = 35$ minutes. The average speed over this interval is

$$\text{Speed} = \frac{\text{distance}}{\text{time}} = \frac{0.90 \times 404 \text{ miles}}{35 \text{ minutes}/60 \text{ minutes per hour}} = 623 \text{ miles per hour}$$

This seems reasonable because a modern jet is limited to subsonic speeds (say, 700 miles per hour) and would cruise at something lower. Just then the travel agent calls back to say that on a brochure the 747 is listed in the 600- to 650-mile-per-hour category.

Going back to the original problem, it appears that of the assumptions made, the times for takeoff and landing are the least accurate. On the other hand, if these stretch out due to weather or traffic the schedule is not maintained. Considering the limits of accuracy of the data, a proper answer would be 600 to 630 miles per hour.

Here is a good opportunity to test your aptitude for one phase of engineering. The above solution has been divided into five separate steps as indicated by the five paragraphs. Look back over the solution and see if you can discover what distinct process was performed in each paragraph. Write down the results of your analysis before reading further.

Now, consider a slightly more involved problem. As you read the solution, try to analyze the work into distinct steps.

Problem B

You are the production engineer of a small manufacturing plant. One morning the purchasing agent brings the news that the company that supplied a certain small shaft for the amount of $2200 per thousand has had a fire and is out of business. The only other supplier wants $3200 per thousand. What is to be done?

Well, you say to yourself after the purchasing agent leaves, we might make them ourselves or maybe it's cheaper to buy them outside. We do have a turret lathe that could do the job, and our shop is not unusually busy. I wonder what it would cost to make the shafts in our own shop taking into account labor, materials, and overhead. (*This first step is an analysis of the overall situation; it involves identifying the important factors and deciding on the general method of approach.*)

I need to know how long it would take for our workers to make the shafts. I ask the turret-lathe operator to set up for production (this takes 2 hours), and he makes five shafts in times of 6, 5, 4, 3, and 3 minutes, respectively. I tell him to finish out the day on the shafts in order to take advantage of the setup time already invested. The industrial engineer tells me that she figures $20 per hour including overhead for shop time on the turret lathe. The shafts are 6 inches long, and the purchasing agent estimates the steel bar stock will cost about $1.20 per foot. If I assume 3 minutes per shaft, specify costs for shop time and material, and neglect all other charges, I have a solvable problem: What is the cost per unit of making shafts on a 3-minute schedule with shop time at $20 per hour and raw materials at 60 cents per shaft? (*The purpose of the second step is to reduce the original vague problem to a question that can be answered.*)

At 3 minutes per shaft, the output will be:

$$\frac{60 \text{ minutes per hour}}{3 \text{ minutes per shaft}} = 20 \text{ shafts per hour}$$

The labor cost will be:

$$\frac{\$20 \text{ per hour}}{20 \text{ shafts per hour}} = \$1.00 \text{ per shaft}$$

The total cost will be:

$$60 \text{ cents material} + \$1.00 \text{ labor} = \$1.60 \text{ per shaft}$$

(*The simplified question is then answered.*)

For 1000 shafts, this would be $1,600. This agrees with the previous price of $2200 per thousand, allowing for profit and costs of transportation and handling. (*The answer to the simplified question is checked by an independent method.*)

In looking over my solution, I feel that the assumptions are quite conservative. The machinist got down to 3 minutes quickly, indicating that the time allowance is liberal. The $20 per hour figure for the turret lathe has been developed after careful study. The $1600 margin between our costs and the outside quotation will cover extra management charges and still yield a good saving. I recommend that we make the shafts in our own shop. (*The final step is application of the result for the simplified problem to the original situation.*)

THE ENGINEERING METHOD

While engineering problems vary greatly in scope and complexity, the same general approach is usually employed. First comes an analysis of the overall situation and a preliminary decision on a plan of attack. In line with this plan, the usually broad and vague problem is then reduced to a more or less straightforward question that can be clearly stated. The stated question is then answered by direct or indirect means. The results for the stated question are always checked for accuracy. Finally the results for the stated question are interpreted in terms of the original problem. The five steps then are *analysis, statement, solution, check,* and *interpretation.*

Preliminary Analysis of the Overall Problem

Engineering problems are never in the form of complete, well-defined questions of the type found in textbooks. The problem handed to the engineer is usually quite vague and lacking in factual data. The electronic development engineer might be asked: How can we obtain stable oscillator operation at 100 MHz? The assignment for a structural design engineer might be merely: Design a two-lane bridge across the Guadalupe Creek at Hedding Street. The refinery operating engineer might be asked: What is wrong with the new cracking unit? In each case, the first step is a preliminary analysis to determine what type of problem it is, what the important factors are, and what approach or approaches appear promising.

The initial step includes determining the nature of the answer. Will it be a magnitude, as the 600 to 630 miles per hour in Problem A? Will it be a recommendation as in Problem B? Will it be merely an answer of "yes" or "no," a quality (corrosion resistant), or a plan (build according to the accompanying drawing)? The form of the answer may indicate what type of problem it is—technical, economic, or personal. Until the nature of the answer has been determined, the assignment is not clearly understood.

Another preliminary consideration is: What are the important factors and which are insignificant? Usually the engineer has available far more information than can be used effectively and, at the same time, there is almost always certain knowledge that is missing that would be helpful. The availability of data concerning the various factors obviously affects the selection of the method to be employed.

Steinmetz is quoted as saying:[1] "The most important part of engineering work—and also of other scientific work—is the determination of the method of attacking the problem." In Problem A, direct, experimental, and analytical approaches were considered; Problem B was solved using a combination of all three. Once a decision has been made on the approach to be followed—and frequently this is where the engineer makes his or her major contribution—then work on the problem can proceed.

Statement of the Question

The second step consists of reducing the original problem to one or more specific questions for which adequate data are obtainable and to which answers can be found. The necessary data may not be right at hand; obtaining the information may require a study of records, a search of the technical literature, consultation with an expert, or some original experimental work.

Frequently, the original problem must be simplified by making idealized assumptions. In Problem A it was recognized that the actual variation of speed throughout the flight was unobtainable, and therefore the assumption was made that takeoff and landing each consumed about 5 minutes. In Problem B, it was decided that the primary factors in determining the cost were material, labor, and overhead, and all other costs were neglected. Each of these assumptions simplified the solution.

Sometimes, a complex problem can be broken down into a series of simple questions. In Problem B the sequence was as follows: How long does it take to make a shaft? What, then, is the labor and overhead charge per shaft? What is the material cost? What is the total cost per shaft? How does this compare with the quoted price? What do I recommend? The more complex the problem, the greater is the need for considering one aspect at a time.

Frequently, it will be found that as the solution progresses, additional assumptions must be made. In other cases, later developments indicate that certain assumptions were unnecessary. Since all important engineering problems are being solved for the first time, no two can be handled in exactly the same way. The procedure must be varied to suit the circumstances.

[1] T. J. Hoover and J. C. L. Fish, "The Engineering Profession," Stanford University Press, Stanford, Calif., 1950.

In any problem, however, this second step should result in a definite statement of a simplified question or questions for which there appears to be adequate information.

Solution of the Simplified Problem

The solution consists in answering the questions asked in the second step. These answers may be obtained by using direct or indirect means.

Direct Approach   The direct approach is very important in engineering; it includes obtaining the answer by direct observation, from records or other sources, or by asking someone who knows. *Measuring* a desired quantity is one direct approach. Many advances in engineering and science have been preceded by improved methods of measurement that revealed new information. In all scientific work it must be remembered that *exact* measurements are not possible; measurements are comparisons that are approximations within certain limits. As these limits are narrowed, the measurements become more *precise*. It should be noted that it is misleading to put qualitative factors on a quantitative scale. For example, it would be deceptive to put style on a 0 to 10 scale and attempt to treat the results mathematically. On the other hand, some qualitative factors have a quantitative basis which may have been overlooked or disregarded. For example, color in the garment industry is a qualitative factor, while color to the scientist is a matter of wavelength, which can be measured with extreme precision.

Estimating is a valuable engineering approach used as a quick and inexpensive method to approximate an answer, to predict results of more precise methods, or to check the results of other methods. Estimating means to form, on the basis of judgment, an opinion from imperfect data. For example, the construction engineer estimates the cost of a project by combining calculated costs with those that can only be predicted. Another type of estimating involves imaginary measurement; the height of a building can be estimated by imagining it as so many times as high as a 6-foot-tall person. In these two examples the opinion is the result of applying judgment developed from previous situations.

Going directly to the result on the basis of experience but without any conscious reasoning is called *intuition*. Intuition is employed many times each day by every one of us. For example, the driver of a passenger car approaching an intersection at the same time as a trucker does not employ a calculator to determine who has the right of way. It is a common experience for a young engineer to present the results of a day's calculations, only to have an experienced superior say: "It just doesn't look right." Intuition based on experience has an important place in engineering. On the other hand, hunches or unjustified intuition are not acceptable.

Indirect Approach   Much of engineering work requires indirect reasoning, and this approach is usually emphasized in engineering schools. Many problems can be solved by a combination of *analysis* and *synthesis*. Analysis consists in breaking something down into its elemental parts. A conscious analysis is usually the first step in understanding a problem; it involves identifying the factors and determining their relationships to the problem and to one another. Referring again to Problem B (third paragraph), the total cost was analyzed into material plus labor charges. The factors affecting labor cost were recognized to be time per shaft and charge per unit of time. What analytical processes took place in Problem A?

Analysis is carried to the point where the elements can be evaluated. Frequently this evaluation is performed by means of *deductive* reasoning. Deduction is reasoning from the general to the specific or applying the general law to the specific case. In Problem A, the desired quantity (speed), was identified as governed by the general law, i.e., distance = speed × time. Substitution of specific values of distance and time permitted the calculation of a specific speed.

In contrast, *inductive* reasoning works from the specific to the general or develops the general law from specific observations. Inductive reasoning is the tool of scientists and research engineers in discovering new laws, while deductive reasoning is more frequently used by engineers in applying general principles to specific problems. Much of engineering training is directed toward providing a background of principles and techniques by means of which the factors can be evaluated after they have been correctly identified.

Synthesis refers to the combination of separate elements into a whole. Synthesis is an important part of design, whether it is the design of a machine, a process, an operating procedure, a report, or a problem solution. After a problem has been analyzed into its components and the components evaluated, the pieces are put back together and the answer built up. In what ways was synthesis used in Problem B?

In both synthesis and analysis, engineers make frequent use of models, representations of reality that are more useful than the real thing. A research engineer, for example, may construct a model of a tunable laser that fits her laboratory observations and indicates the direction for new experiments. A systems engineer may employ a mathematical model to represent a complex control unit and show it as a labeled block diagram. Because of their importance in problem solving, modeling and design are described in some detail in later sections.

Checking the Answer to the Simplified Problem

There are no answer books in engineering practice; usually engineers who do the work are the only ones who can judge the correctness of their results.

Because their recommendations may become the basis for large investments in money, time, or human life, and because their professional reputations ride on each decision, the engineers must learn to check their own work effectively.[2] In an engineering problem, the method of solution must be checked as well as the calculations. Manipulative mistakes and numerical blunders can be corrected by careful repetition of individual steps. The best check on a method is solution by an independent method.

Methods of Checking   Any method of solution constitutes a possible check for another method. In most cases the ultimate test is a practical application; the other methods are used as quicker or less-expensive substitutes for the real thing or as preliminary steps prior to the actual construction or operation. An experimental solution or check may be obtained in a laboratory by means of a test carefully contrived to yield the desired results. Aircraft-design problems are usually checked experimentally in wind tunnels. In some situations the accuracy of the experimental check can be obtained at less expense by using models and extrapolating the data to the full-scale unit. For example, a 4-foot scale model may be used to represent an aircraft with a 40-foot wingspread, or an electrical circuit may be used as a model (analog) of a heat-transfer problem. The analytical method depends on a knowledge of fundamental principles derived from previous practice or experimental work and has the advantage of great flexibility in the type of problem attacked. Estimating has already been discussed as a method of quick solution, and it has limitations, but the advantage in speed and the generally high dependability justify its frequent use in engineering.

In Problem B the resultant cost of $1.60 per shaft was checked by comparison with the previous price per thousand on the basis of experience. What two checks were made in Problem A?

The Time for Checking   Checks should be applied throughout the solution rather than reserved until the end. Frequent checks increase efficiency by avoiding the waste of time caused by mistakes made in the early stages. Frequently, a check solution by estimating or intuition should be made before beginning the more detailed solution. This procedure is especially recommended for engineering students. After the preliminary analysis when the nature of the answer has become clear, make a guess as to the result. This intuitive solution will provide a more reliable check if made ahead of time than if you wait until the analytical results are available to influence your thinking. In any solution, you must guard against being so prejudiced in favor of your own work that you slight the checking process.

---

[2] As someone has said: "To err is human; to check is engineering."

Interpretation of the Result

Having solved a simplified problem and checked the results, you must now go back and interpret the results in terms of the original problem. If assumptions were made to simplify the solution, their effect should now be considered. If certain factors were neglected, they should now be taken into account by either modifying the result or indicating the precision of the answer. In many cases the problem actually solved was for an idealized situation, and the result must be modified to fit the practical case.

The form of the answer may not agree with the original assignment. Problem B yielded a result in dollars, while the original question called for a recommendation. Sometimes the result will be in a general form to be applied to a specific case. Other times the result will be specific but will have general implications that should be pointed out.

In any case the answer must be reported back to the person making the assignment. This presentation may be short or long, formal or informal, oral or written, one page or several volumes in length. The form of the interpretation must be governed by the requirements of the reader and the intended use of the report. Your instructor will be interested in your method and reasoning; a client will be interested in an answer in dollars and cents; another engineer will wish to know your assumptions and where you used your own judgment. In many situations, you may spend more time on the interpretation than on all the rest of the solution.

MODELING

A model is a useful representation of reality, useful in that it is helpful in analyzing behavior or predicting performance. For example, the Bohr model of the atom (Figure 12-1) consists of a massive nucleus carrying a positive charge surrounded by one or more orbital electrons carrying negative charges. The Bohr atom is useful in explaining atomic spectra in terms of

FIGURE 12.1
The Bohr model of atomic structure.

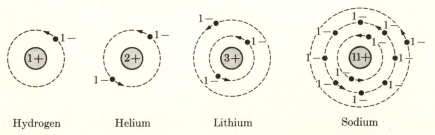

Hydrogen   Helium   Lithium   Sodium

transitions of electrons from one orbit to another. For other purposes it has been replaced by the more sophisticated quantum mechanical model.

Modeling is one of the common activities of engineers. A model may be useful to the engineer because it is simpler, smaller, cheaper, or easier to handle than the real phenomenon, device, or system it represents. Engineering models range from the simple expression relating force to displacement in a spring to the complex representation of a complete urban transportation system. The principal requirement is that the model retain the essential characteristics of the real thing. Of particular importance to engineers are physical, mathematical, and circuit models and computer simulations.

Physical Models

The airfoil section used in wind-tunnel studies of lift and the pilot plant used in development of chemical processes are physical models. They represent reality but they are simpler and smaller and therefore less expensive than the real thing. In preparing physical models, only the important characteristics are retained. For example, an aerodynamic model might be machined from solid cast iron, whereas the real section would be made of hollow titanium;

After analytical and computer models are designed and studied, small-scale models are frequently built to further verify performance and reliability prior to prototype construction. Here a model of the space shuttle is being installed in a transonic wind tunnel at the Arnold Engineering Development Center. (*U.S. Air Force*)

only the shape is important. Extrapolations from observations on small-scale physical models must be based on knowledge of the way the behavior of interest varies with dimension; simple proportionalities seldom apply.

Mathematical Models

In many cases, an equation or a set of equations can be used to represent the behavior of a device or system within acceptable limits. The major advantage of mathematical representation is that a great variety of tools and techniques are available for manipulation and interpretation once the models are formulated.

The derivation of a mathematical model is illustrated in Figure 12-2. The heavy line indicates the actual observed relation between force and displacement of a given spring. Over a limited range, the force is directly proportional to displacement and the characteristics can be approximated by $F_S = Kx$, where K is a constant of proportionality. We say that $F_S = Kx$ is the *mathematical model* of the spring. (At higher displacements $F_S = K_1x - K_2x^2$ would be a more accurate model.) Similarly, $F_M = Ma$ is a mathematical modeling of Newton's second law describing the behavior of mass M undergoing an acceleration a due to force F (Figure 12-3). The viscous damper, a piston sliding in a cylinder filled with oil, develops a force F_D that is approximately proportional to velocity v; its model is $F_D = Dv$ where D is the damping coefficient. The mathematical model for the spring-mass-damper

FIGURE 12.2
Force-displacement characteristic of a spring.

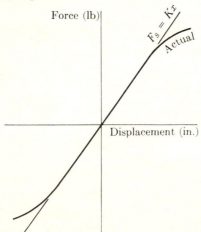

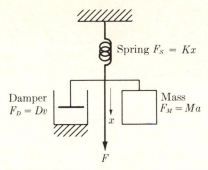

FIGURE 12.3
A spring-mass-damper system.

system is obtained by applying the principle that the applied force F is just equal to the sum of the retarding forces, or

$$F = F_M + F_D + F_S = Ma + Dv + Kx \tag{12-1}$$

Computer Simulation

In a pilot trainer, a trainee can experience all the sensations of flying without leaving the ground. If he opens the throttle, on the screen in front of him the runway appears to be rushing by; if he pulls back on the stick, the horizon disappears and a climb is simulated. The same type of *simulation model* is used to train astronauts for activities on the moon. In a similar way, a skilled pilot can "fly" a proposed aircraft design and evaluate its performance and handling before any part of the aircraft has been built. The large digital computer is ideal for simulating complex systems in which there are many variables. The computer can be programmed to vary the design parameters in a prescribed way in order to optimize certain performance characteristics.

Simpler systems can be simulated by the analog computer in which certain variables, usually voltages, behave in just the same way as the given system variables. Consider, for example, the spring-mass-damper system shown in Figure 12.3 and defined by Equation (12-1). Solving Equation (12.1) for acceleration,

$$a = \frac{F}{M} - \frac{D}{M} v - \frac{K}{M} x \tag{12-2}$$

Using calculus notation, velocity is the integral over time of acceleration or $v = \int a \, dt$, and displacement is the integral over time of velocity or $x = \int v \, dt$. An analog computer consists of *summers* (Σ), *integrators* ($\int$), and *multi-*

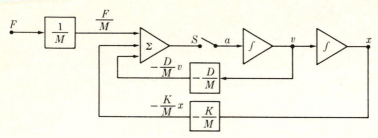

FIGURE 12.4
Analog computer simulation of a spring-mass-damper system.

pliers conveniently arranged to solve equations. In Figure 12.4, the variables *F*, *a*, *v*, and *x* are represented by voltages. Closing switch S forces the system to simulate Equation (12-2); for any applied force, the corresponding acceleration, velocity, and displacement can be displayed as functions of time on an oscilloscope.

Circuit Models

A *circuit* is a system in which the variables are defined in terms of a single dimension—the position along a line. The flow of fluid in a hydraulic system or the flow of traffic in a highway system can be described in circuit terms. Working primarily with electrical circuits, engineers have derived a great variety of analytical techniques that become available once a device or system has been represented by a circuit model.

Consider, for example, the transistor shown in Figure 12-5a. Two *n*-type semiconductors (negative carriers) have been joined to a *p*-type *base* (positive carriers) to form *alloy junctions* between the base and the *emitter* and the *collector*. Such a transistor is represented symbolically as shown in the

FIGURE 12.5
An NPN transistor.

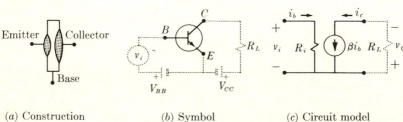

(*a*) Construction   (*b*) Symbol   (*c*) Circuit model

elementary amplifier circuit of Figure 12-5b; the arrowhead that identifies the emitter terminal always points toward n-type material. In predicting the performance of this amplifier, the transistor is replaced by the simple circuit model shown in Figure 12-5c.

The characteristics of the transistor input and output are shown in Figure 12-6. In an amplifier, we are interested in small variations or *signals*. The input characteristic shows that a small variation in emitter-base voltage ΔV_{BE} is accompanied by a small variation in base current ΔI_B. The ratio of voltage change to current change defines an *input resistance* R_i. The output characteristic shows that a small change in base current, ΔI_B, results in a small change in emitter current, ΔI_C. The ratio $\Delta I_C/I_B$ is called the *current gain* β.

In the circuit model of Figure 12-5c, the signals are shown by lowercase letters; the dc supply voltages V_{BB} and V_{CC} do not appear on the small-signal model. Using Ohm's law, the base current is determined by the input voltage or $i_b = v_i/R_i$. The collector current is just equal to the current gain times the base current or $i_c = \beta i_b$. In this case, $\beta = \Delta I_C/\Delta I_B \cong 1.0/0.02 = 50$. A typical value of R_i is 1000 ohms. For a load resistance $R_L = 2000$ ohms, the *voltage gain* of the amplifier would be

$$G_V = \frac{v_o}{v_i} = \frac{i_c R_L}{v_i} = \frac{\beta i_b R_L}{i_b R_i} = \frac{\beta R_L}{R_i} = \frac{50 \times 2000}{1000} = 100$$

or the output voltage is 100 times the input voltage. Once the circuit model for this transistor has been determined, it can be used to predict the performance of the transistor in any circuit.

FIGURE 12.6
Transistor characteristics.

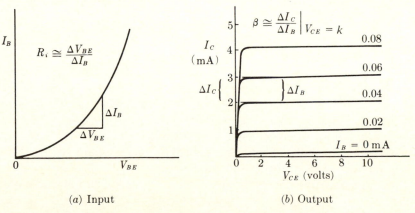

(a) Input

(b) Output

DESIGN

The problems discussed in this book require analysis of an existing situation. Of great importance in engineering is the process of creative synthesis or design—putting ideas together to create a new solution that will solve the problem in the best possible way. In Problem B, for example, the engineer's first thoughts might be: What other methods of fabricating the shaft are available? Could it be built up out of stock rod and sleeves? Do we need that shaft after all? Is there a better way of doing the job?

Although design involves the same basic steps as general problem solving, there are some distinguishing differences. One of these is the emphasis on synthesis, generating new ideas, and combinations of ideas. Another difference is that whereas problem solving was pictured as a straight-line process, design is characterized by *iteration,* repetition of earlier steps in light of new information. Because there are many solutions to a given design problem, several possible approaches must be carried along until the best available approach can be identified. Then that approach must be optimized by maximizing performance or minimizing cost or meeting a time limitation. Finally, the interpretation of the solution to a design problem is in the form of a detailed design and set of specifications that incorporates all the decisions made along the way. (Even then the design is not finished; experience on the production line or in the field provides additional feedback and initiates another iteration.)

The complexity of the design process is illustrated in Figure 12-7. After defining the need (step 1), which may be quite different from the perceived need, the design engineer begins to generate ideas in the form of possible approaches and solutions. At the same time, he defines the boundary conditions by stating the apparent limitations and specifies the criteria by which proposals are to be evaluated.

The next stage is the so-called feasibility study in which he identifies proposals that appear to satisfy the criteria without exceeding the limitations. (The review of possible approaches may result in reassessments of criteria and limitations that are shown as information feedback on the diagram.) The selection of feasible approaches focuses attention on critical problems or bottlenecks and leads to a series of problem statements—specific questions (step 2) regarding components or aspects of the various proposals, usually formulated in terms of models describing the system and its environment. (Formulation of these questions may generate new ideas that are introduced into the design process.) In answering the specific questions (step 3), the engineer creates or synthesizes new devices or processes or subsystems that will satisfy the basic need or solve the original problem. (Again there is feedback that may modify criteria and limitations.)

In a design problem, checking (step 4) includes testing and evaluation in

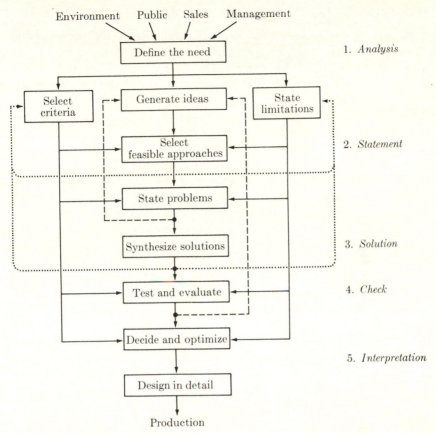

FIGURE 12.7
The design process as an example of problem solving.

terms of the specified criteria and limitations as well as checking of the calculations and the models on which they are based. The testing may involve laboratory experimentation, testing of full-scale prototypes, or computer simulation. (The knowledge gained from experiment may result in new ideas for consideration, and another iteration begins.) At some point the engineer must select the best concept, optimize its characteristics, and make those difficult decisions on which the final design is based (step 5). In detail design, all the remaining technical questions are answered, dimensions and tolerances are stated, materials and finishes are specified, and methods of assembly and testing are prescribed. The design is finished—until the first reports begin to come in from production and sales engineers.

APPLICATION OF THE ENGINEERING METHOD

Let us now return to the original problem and solve it by the suggested approach.

Problem

Assume that you are a highway engineer concerned with the problem of runaway trucks on a long downhill stretch of highway. What could be done to reduce the chance of such an occurrence?

1 Preliminary Analysis At the beginning of a long downgrade it is customary to post a sign saying: "Trucks use low gear." In low gear, the potential energy of the truck at the top of the grade is safely absorbed in compressing air in the engine. Sometimes, however, a driver fails to get the truck in low gear. Because brakes are not designed to absorb and dissipate energy on a continuous basis, some other method is needed for absorbing the kinetic energy of a fast-moving vehicle. It must be reasonable in cost because it will be used fairly seldom, and it must not introduce some new hazard. At this time a preliminary design to test feasibility would be appropriate.

2 Statement One possibility would be for the driver to use a median dividing strip in some way, but that opens the possibility of a head-on collision with an approaching vehicle. Another possibility is to use a center dividing rail by snubbing the front tires against it; this is frequently used, but the maneuver takes some skill and the truck must cross the fast lane to get into position. Still another possibility would be a safety exit ramp. The preliminary design would consist of specifying the length, the grade, and the surface material, which depend on the kinetic energy to be absorbed—a function of the maximum speed of the runaway vehicle.

Pursuing the third approach, we make the following simplifying assumptions:

The design speed is 100 miles per hour (147 feet per second).
The exit ramp is on a 15 percent uphill grade.
The surface is covered with loose sand.
The retarding effects of mechanical friction and air resistance are negligible.

On the basis of these assumptions, we formulate a straightforward problem: Calculate the length of ramp required to bring a truck to rest under these conditions.

3 Solution Our model of the truck-ramp system (Figure 12-8) is a free-body diagram showing all the forces acting on the truck. The weight W acts

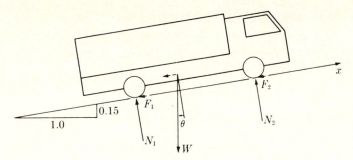

FIGURE 12.8
Free-body diagram of truck slowed by friction.

downward and the normal forces N_1 and N_2 act perpendicular to the roadway surface. The retarding forces are due to friction (F_1 and F_2) and to a component of weight along the slope (W_X).

The sum of the normal forces must be equal and opposite to the normal component of the weight, or

$$N = N_1 + N_2 = W \cos \theta = W \cos (\tan^{-1} 0.15) \cong 0.99 \, W \cong W$$

Where the coefficient of friction $f = 0.20$ for a tire rolling in sand (obtained from an engineering handbook), the total friction force is

$$F_1 + F_2 = fN_1 + fN_2 = f(N_1 + N_2) \cong 0.20W$$

The component of weight along the slope is $W_x = W \sin (\tan^{-1} 0.15) = 0.15W$, and the total retarding force is

$$F = F_1 + F_2 + W_x = 0.2W + 0.15W = 0.35W$$

Since force = mass × acceleration, this force produces a deceleration

$$a = \frac{F}{M} = \frac{Fg}{W} = \frac{0.35Wg}{W} = 0.35g = 11.2 \text{ ft/s}^2$$

We know that the velocity v is related to the distance traveled s by the relation $v^2 = 2as$. To reduce the velocity from 147 feet per second to zero,

$$s = \frac{v^2}{2a} = \frac{(147)^2}{2 \times 11.2} = 960 \text{ ft}$$

4 **Check** For a check, we can approach the problem from a different direction. We know that some of the total kinetic energy (KE) will be dissipated in friction and the rest will be converted to potential energy (PE). Therefore,

Friction loss $=$ force $\times$ distance $= F_S = 0.2W \times 960 = 192W$

PE gained $=$ weight $\times$ height $= Wh = W \times 960 \times 0.15 = 144W$

Initial KE $= \frac{1}{2} Mv^2 = \frac{1}{2} (W/32.2)(147)^2 = 336W$

which checks.

5 **Interpretation** Looking again at the assumptions, we note that mechanical and air resistance will actually help. On the other hand, the assumed maximum speed may be exceeded in the future and the friction characteristics of the sand may deteriorate with time or weather. For the preliminary design of a safety exit ramp, we will specify: one-quarter mile (1320 feet) of 15 percent grade with a loose sand surface.

SUMMARY

One of the characteristic activities of engineers in all functions is the solution of problems. A logical, step-by-step approach is recommended. The major steps are as follows:

1 Preliminary analysis of the overall problem
2 Statement of the question in simplified form
3 Solution of the simplified problem
4 Check on the answer to the simplified problem
5 Interpretation of the result in terms of the original problem

In college engineering courses, the major emphasis is on step 3 with some training in the background required for steps 2 and 4. Experience on the job will help in steps 1 and 5 particularly. Imagination and creativity are required in step 3. As you work simplified problems in specialized courses, bear in mind that, in practice, problems do not come wrapped up in such neat packages.

REFERENCES

1 Beakley, G. C., and H. W. Leach, *Engineering: An Introduction to a Creative Profession*, 4th ed., Macmillan, New York, 1982.
2 Eide, A. R. et al., *Engineering Fundamentals and Problem Solving*, McGraw-Hill, New York, 1979.

3 Krick, E. V., *An Introduction to Engineering: Methods, Concepts, and Issues,* Wiley, New York, 1976.

4 Polya, George, *Mathematical Discovery; On Understanding, Learning, and Teaching Problem Solving,* Wiley, New York, 1967–68.

5 Rubinstein, Moshe F., *Patterns of Problem Solving,* Prentice-Hall, Englewood Cliffs, N. J., 1975.

ASSIGNMENTS

Solve problems 1 to 5 using the approach outlined in this chapter. Label clearly the five steps in the solution. State all assumptions made; your instructor will discuss the engineering principles involved and assist you in making the simplifying assumptions.

1 Two roommates are going to drive a 1978 Mustang from New York to San Francisco at the close of school. They plan to share all transportation expenses, being fair to the one who owns the car. Estimate the total cost for the trip.

2 A dormitory is located on a 200- × 300-foot corner lot. The dormitory and landscaped area are to occupy the 100- × 100-foot area at the corner; the remaining area is available for student parking. How many cars can be parked?

3 What minimum horsepower is required for a 3200-pound automobile to enter an expressway during a high-speed, high-density rush period? Repeat for a 1600-pound automobile.

4 A 3-ton load is to be supported by two timbers bolted together at the top and bolted to the floor as shown in Figure 12.9. What size timbers should be used?

5 A granular product weighing 0.25 pound per cubic inch is to be shipped in cylindrical tin cans containing 20 pounds of product. Design an optimum can.

6 Design the hydraulic system for a service-station car lift.

7 An open-top steel tank 6 feet high and 6 feet in diameter costs $350. The ⅛-in plate costs $150 and the fabrication (primarily welding) costs $200. Estimate the cost of a similar tank to hold four times as much.

8 An ore hoist in a 1150-foot-deep potash mine handles skips with an ore capacity of 8.4 tons. The two skips weigh 6 tons each but counterbalance each other with

FIGURE 12.9
Timber support.

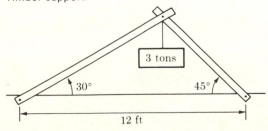

one loading while the other dumps. Top speed of the $1^5/_8$-inch wire rope is 1500 feet per min. Skips are accelerated for 10 seconds. The dumping and loading operation is scheduled for 16 seconds. What is the hoist capacity in tons per hour and what power do you recommend to operate it?

9 From a Peace Corps volunteer working with the Kashmir Tourist Agency, VITA (Volunteers for International Technical Assistance) has received a request for a rope ski tow design. The hill is about 120 feet high and the slope is 700 feet long. The local electric voltage fluctuates widely, but jeeps could be used for power. What would you suggest?

10 In the disk brakes of the Ford GT MK-II, an iron-molybdenum alloy rotor is rubbed by caliper friction pads. At LeMans, the winning car must average over 125 miles per hour and rotor temperatures of 820°C are expected. Estimate the amount of kinetic energy that must be dissipated in a braking turn like the Mulsanne corner.

11 The weight of a single-stage rocket is limited to 40,000 pounds. When kerosene burns with liquid oxygen at the rate of 1 pound per second, it develops a thrust of 250 pounds. What final vertical velocity can be expected from this rocket using this fuel?

12 Compare a 20-gallon gasoline tank and a set of lead-acid storage batteries occupying the same volume on each of the following criteria: total energy content, mechanical energy available for automobile propulsion, and energy cost per mile at in-town speeds.

13 A deluxe mountain cabin uses electricity for lighting and cooking and propane gas for heating water for a family of two adults and four teenagers. How big a storage tank is required if gas is delivered every two weeks?

14 In a large power plant the steam input to the turbine is at 1000°F and the condenser cooling water from a lake is at 50°F. The overall plant efficiency is 40 percent. If improved engineering would permit using steam at 1100°F, what overall efficiency would be expected? If a thermionic converter with cathode at 3000°F and anode at 2000°F is added ahead of the steam turbine, what overall efficiency might be expected?

15 A certain line of incandescent lamps is designed to last 1000 hours at 115 volts. Experiment shows that if the lamps are operated at 120 volts, they only last about 600 hours. Your boss asks you: "For those hard-to-reach fixtures in the dome of the warehouse, how about operating 115-volt lamps at 110 volts?"

16 A 24-volt battery mounted on a handtruck supplies power to an aircraft starter (total armature circuit resistance = 0.1 ohm) through a pair of No. 10 copper wires (1 ohm per 1000 feet). The open-circuit voltage of the battery actually measures 28.0 volts; when the battery is momentarily shorted through an ammeter, a current of 1400 amperes flows. Derive a circuit model for the battery and predict the power supplied to the starter at the instant the starting switch is closed.

17 A vibration sensor with an internal resistance of 5000 ohms develops an open-circuit output voltage of 1 millivolt. To drive the available recorder requires 25 milliwatts into an input resistance of 1000 ohms. Design an amplifier (a preliminary design) to provide the necessary voltage gain using the available transistors ($\beta = 30$ and $R_i = 2000$ ohms).

18 Your old aircompressor unit has an overall efficiency of 50 percent and a salvage value of $400; the annual power bill is $3000. At a cost of $3000, you can obtain a replacement unit with an overall efficiency of 70 percent. If money is worth about 8 percent and investments of this type should pay off in five years, what do you recommend?

19 You are planning to buy a new computer for $200,000. You estimate that annual maintenance and taxes will cost $5000 per year. After a few years, a computer of this type drops in value to about one-half its original price. A computer leasing agency offers to lease you a similar computer and pay all maintenance costs for $50,000 per year. Which do you recommend?

13

Energy
Environmental pollution
Space
Productivity

CHALLENGES
TO ENGINEERING

As an engineer you will be part of the technical future of our country; it will be an interesting, rewarding and, at times, frustrating experience. It will be interesting because you may be part of the technical team that develops a new technology such as the synthetic fuels that will bring a higher standard of living to people in every nation in the world. It will be rewarding because you will gain a sense of satisfaction in knowing the world is a better place in which to live because you have contributed to the well being of its citizens.

However, your job may also be frustrating! Fifty years ago purely technical solutions were readily accepted by society; it is not so today. Today, engineering proposals must be explained and justified and are frequently debated in the public sector before decisions are made. This chapter will discuss some of the engineering challenges that lie ahead. You must understand the issues and be prepared to discuss their implications not only with other engineers but with the general public. Technical solutions are available to many of the problems that face society today, but a social consensus must be reached before work can start. You must understand not only technical aspects but also social concerns to be a creative problem solver, an effective decision maker, and an articulate representative of your profession.

ENERGY

While waiting in line at a service station, a young engineer looks quickly through the evening paper. Headlines proclaim: "Congress Works on Legislation for Energy Independence"; another reads, "The U. S. has Adequate Energy for Next 200 Years." Like many of us, she wonders: If the U. S. has all the energy it needs for 200 years, why is Congress working on energy independence and why is she waiting in line for gasoline which seems more expensive every time she buys it? This section will briefly outline the approaching energy crisis which the American people and their elected representatives don't seem to understand or to be able to cope with. The types of energy available to correct this problem will be discussed along with their advantages and disadvantages.

Importance

If new products are to be competitive in the marketplace and if an industry is to have a future in our changing technical society, the engineer must take into account the supply of energy.

As shown in Figure 13-1a our standard of living, which is reflected in gross national product (GNP) and employment (jobs), is directly related to energy consumption. As energy becomes more expensive it will be used more carefully but it will still be a major requirement for our high-technology industry and the quality of life we enjoy. Figure 13-1b compares energy consumption in the developed and undeveloped nations. According to this projection, the United States will increase its per capita energy consumption only slightly in the decades ahead but developing nations will need to greatly increase their consumption to attain the higher standard of living they seek. In fact, they must almost quadruple the present per capita energy consumption, and that is a most difficult challenge.

Liquid Petroleum

The problem is not that there is a shortage of energy but rather that the world depends heavily on only one kind—liquid petroleum. The United States, for example, is presently importing about 50 percent of the liquid petroleum it needs. As seen in Figure 13-2, most of the world's supply of oil and gas reserves lies outside of the United States. The Communist countries consume all they produce and will soon need additional supplies. Most of the

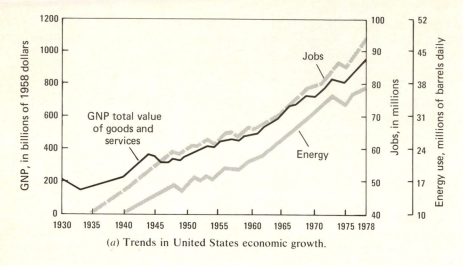

(*a*) Trends in United States economic growth.

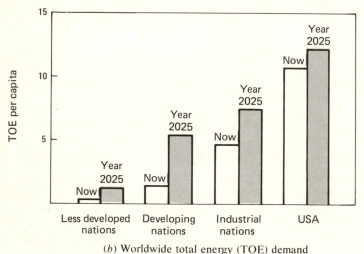

(*b*) Worldwide total energy (TOE) demand

FIGURE 13.1
Energy use and economic growth. (Andre Giraud, *World Energy Resources*, 1976.)

remainder of the world's supply is controlled by the thirteen-member Organization of Petroleum Exporting Countries (OPEC). So, OPEC has a monopoly on a large percent of the world's available oil supply. From 1973 to 1980 this monopoly resulted in a rapid escalation in prices, as well as an availability dependent primarily on the political stability and political interests of the Arab world. Further complicating the problem is the fact that

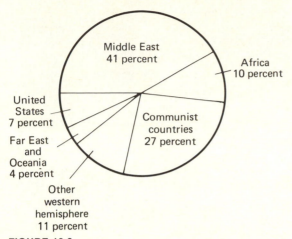

FIGURE 13.2
World oil and gas reserves. (*Central Intelligence Agency.*)

world resources of petroleum are rapidly being depleted. Even without a monopoly to contend with, the world will have to change over the next few decades to other types of energy.

Primary Energy Sources

As shown in Figure 13-3*a*, wood was the major fuel used in the United States in the early 1800s. In the late 1800s, coal became our major fuel. In the 1940s petroleum and natural gas replaced coal. Today we are in the process of finding a new major fuel to replace petroleum.

One of the major conclusions of a report on energy-supply projections by the ASCE National Energy Committee, which has been verified by many other national studies, is that the only major sources of energy available until at least the year 2000 are coal and nuclear. [3]

Projections for the Future

Figure 13-3*b* shows the dependence of various parts of our society on oil, or liquid petroleum. Based on present trends, the impact on our economy of the liquid petroleum monopoly and dwindling supply will get progressively more serious.

Figure 13-4 displays projections of the energy needs of the United States. The energy in the various fuels is expressed in terms of quads. One quad is equal to the energy of $7^1/_2$ billion gallons of gasoline (enough gas to run 10

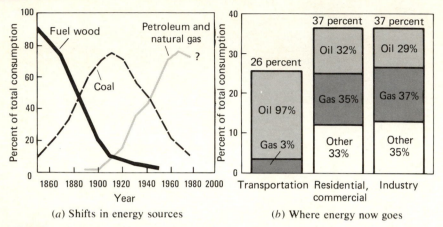

FIGURE 13.3
Energy sources and consumers in the United States [1].

million automobiles for a year) or 46 million tons of coal (enough to fill a string of railroad cars that would extend across the country and back). As shown in Figure 13-4, the most conservative estimate of our energy needs by the year 2000 (110 quads) versus the most optimistic projection of our

FIGURE 13-4
Estimates of United States energy needs and sources of supply [2].

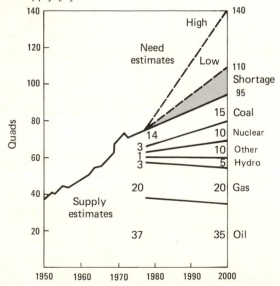

available energy production (95 quads) leaves a shortage of 15 quads. Recalling that a one-quad shortage of energy in 1973 caused the most serious economic recession this country had experienced since the great depression of the 1930s, we can conclude that a fifteen quad shortfall would be devastating.

Predicted changes in available energy sources to cope with the approaching energy crises are also shown in Figure 13-4. Aside from coal, nuclear, and gas, you will notice that the only other domestic energy source large enough to be measured in quads is hydropower. However, it has small potential for growth due to the limited number of dam sites. Hydropower, as well as free solar energy, free wind power, and conservation will all help to lessen the energy problem but cannot eliminate the need for a major new energy supply to replace liquid petroleum. The only way this crisis can be avoided is to undertake a massive increase in production of coal-fired elec-

The Gibson power station is located on the Wabash River near Princeton, Indiana. This is the largest coal-fired plant in Indiana (2,600,000 kilowatts) with another 650,000-kilowatt unit under construction. Power plants are located near large bodies of water for cooling purposes. (*Public Service of Indiana*)

trical plants or of nuclear plants or a combination of both [2]. The longer we wait to start, the worse the energy shortage will impact on our society.

Coal

As shown in Figure 13-5, in terms of coal reserves the United States is in a much more favorable position. As quoted by Dr. Arthur Bueche of General Electric during a speech in 1969:

> If we're going to try to do the job with coal, we must start—right now—to map out and dig 300–400 huge new coal mines, which also means starting—right now—an unprecedented expansion of our national transportation system: hundreds of thousands of new railway locomotives, cars, and barges; new and improved rail rights-of-way; waterways; and coal-slurry pipelines. And, of course, we must build hundreds of new billion-dollar plants to convert this coal into electricity and other forms of energy cleanly and efficiently. We need an all-out effort to create synthetic fuels from coal that will match the job done during World War II to create synthetic rubber when our imports of natural rubber were cut off.

Nuclear

To meet our future energy needs with nuclear power would require a massive commitment similar to that required for coal. Like coal, large uranium reserves are available in this country, as shown in Figure 13-6. However, even though nuclear may be our cheapest, safest, and cleanest energy source, society is reluctant to choose it to solve our energy needs.

FIGURE 13.5
World coal reserves [1].

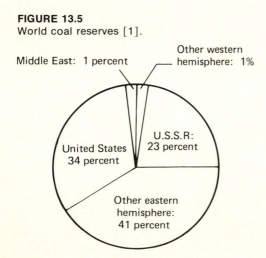

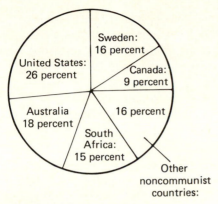

FIGURE 13.6
Uranium reserves of noncommunist nations. (*U.S. Bureau of Mines.*)

Zion Station (nuclear) is on the shore of Lake Michigan. Nuclear energy is a good engineering solution to the energy problem, but whether it will be acceptable to society remains to be seen. (*Jim Hedrich, Hedrich-Blessing*)

This reluctance was exemplified by the Three Mile Island nuclear accident in March 1979. In the public's mind it was an impossible accident; as far as the nuclear industry and the scientific community were concerned, it was unlikely, but not impossible. This accident brought to light the lack of preparation by the industry, the government, and the public to react to a nuclear accident. Less publicized but of equal importance was the low level of risk associated with this accident. The hazard to the public was so small it was not measurable. Considering that 2 million people lived within 50 miles of the Three Mile Island nuclear plant, and that statistically one in six people in the United States die from cancer (330,000 in the Three Mile Island area), the possibility of there being one additional cancer death was certainly not a high level of risk[4]. However, while the American people are willing to take risks with their own lives (like crossing the street during rush hours) they want and expect a much higher level of safety for risks that are imposed on them.

Summary

As an engineer you will be heavily involved in finding solutions to the energy problem. As a student, you should learn the extent of the problem and what alternatives exist; people will expect you to know. Thousands of engineers will be involved when the move to construct the much needed energy plants gets underway. You may be one of them. Even if you are not directly involved, whatever product you are concerned with will need to be built in the most energy efficient way to be competitive. The future of our technical society will depend on the technical competence of you and your classmates and how efficiently we use energy.

POLLUTION

Everybody's for a clean environment, clean rivers, and clean air. And I'm for that. Strongly for that. But when you walk over to that switch and your light doesn't come on, and when you turn on the thermostat and the heat doesn't come up, or the air conditioner doesn't go down, well-then is it that important that we have absolutely pure water and pure air, when we don't have any jobs for our people. . . no money to buy food with? (Gov. James B. Edwards of South Carolina, 1977)[5]

There is a lot of concern about the quality of our environment. Highly qualified scientists disagree on the environmental effects of large increases in coal-fired energy plants to meet our future energy needs. Some feel coal is the best and safest fuel. Others disagree. For example, a Harvard Energy and Environmental Policy Center study reports that 53,000 Americans are dying each year from industrial pollution. An Ohio River basin energy study, conducted by 100 scientists over 4 years at a cost of $4.3 million concluded that by the year 2000:

1 At least 109,000 people living in the Ohio River valley will die from pollution-related diseases primarily linked to breathing disorders linked to air pollution from coal-fired power plants.

2 Crop damage losses could equal $7 billion due to the increases in pollution levels of nitrogen dioxide in the emissions from power plants, which interfere with the photosynthesis process.

3 Acid rain is destroying aquatic life in the eastern United States and Canada and decreasing timber harvests.

To make matters worse, Dr. Robert Jastrow, director of the NASA Goddard Institute for Space Studies, predicts that massive increases in carbon dioxide which would result from a switch to coal-based energy will have severe effects on world climate. Specifically, he predicts the resulting greenhouse effect will melt the west Antarctic ice cap, raise the oceans 15 feet, and result in massive flooding of sea level communities[6]. There are certainly questions in the scientific community about the safety of a massive increase in coal-fired plants to meet our future energy needs.

The nuclear industry also has contributed to environmental concerns, real and imagined, which have been widely reported in the news and exemplified by the Three Mile Island accident. While other countries are rapidly developing atomic and breeder reactors, there is great reluctance by the American people to resort to nuclear power. Clearly, pollution of our environment is an important concern of society. This is also an important concern of engineers who build the plants and devices that contribute to the problem and, it is hoped, to the solution.

As discussed earlier, modern society requires energy. Coal-fired and nuclear energy plants seem to be the only answer to the largest part of this energy need. An energy solution will not be possible unless there exist coal-fired and nuclear power plants. One can readily see that the engineer has a great opportunity to use his creativity along with his professional training to design an energy base that will be acceptable to the American public.

Trade-offs

A solution must consider trade-offs. Julian Carroll, a recent governor of Kentucky, said that one of the more difficult challenges is establishing a "more effective way to balance costs and benefits" in environmental regulations:

> If you plot the cost of cleaning up manufacturing plant discharges versus the percentage of pollutants that you get out of the effluent, there's no question that you reach a point of very small return for a very large investment . . . . In other words, it may be reasonable to get the first 95 percent of the pollutants out, but the cost of

A recent addition to an "integrated gasification/combined cycle" (IGCC) research facility, is a special cleanup system for removing sulfur compounds from the synthetic coal gas. This system is an adaptation of a design long employed for sulphur removal in the chemical process industry. It removes more than 90 percent of the sulfur compounds from the coal gas. "Centerpiece" of the IGCC research facility is an experimental gasifier capable of converting up to a ton of coal per hour into fuel gas. Fuel produced by the gasifier is passed through the cleanup system and then burned in combustion facilities that simulate key elements of a combined-cycle, power-generating installation. (*General Electric Company*)

getting that last 5 percent may be three times as much as the cost of getting out the first 95 percent. And I think when you get to that point, you have to stop and ask whether it makes sense to go further.[5]

How much is enough pollution control? This is not a decision that you or your company can make alone. It is a decision that society is making—frequently on inadequate or inaccurate knowledge. For society to make an informed judgment balancing health and jobs (industrial competitiveness), you and other engineers must understand the problem and contribute in a responsible way to the discussion.

SPACE

Space is not "the final frontier," as a television program tells us, but rather the next frontier. There will always be frontiers for people with curiosity and imagination. It is easy to look back at previous frontiers, such as the Louisiana Purchase and Alaska, and see the role that engineers played in their exploration and development. Do you have the foresight and imagination to look ahead and see the role engineers will play as man moves out into our solar system? This section will suggest a few of these ways.

Energy

What does space contain that interests people and nations? How about energy? A recent Forbes article states: "To satisfy the energy needs of the United States in conventional ways, our electrical utility industry now figures it will have to spend some $800 billion during the next 25 years."[8] Could we avoid the difficulties inherent in the use of coal or nuclear power by developing extraterrestrial sources? Far less than 1 percent of the sun's energy falls on the earth. The rest is lost in space. Would the American people prefer satellite solar power stations? Is this a feasible option? If scientists could place solar collectors in space and beam this energy safely back to earth a limitless nonpolluting form of energy would be available. Preliminary engineering studies have shown that it is technically possible and economically feasible. Should this energy option be kept open for the American people?

Sixty satellites, each producing 5 gigawatts (5 billion watts), could furnish 15 to 20 percent of the electrical needs for the year 2030. However, the developmental cost to build one 0.5-gigawatt pilot plant is estimated to be $35 billion. An additional $50 billion will be needed to build the first 5-gigawatt unit; this includes the cost of a complete space construction complex and a fleet of heavy-lift launch vehicles. Follow-on units would cost only $12 billion each! The cost of this power to the utilities would be 6 cents per kilowatt-hour which would be economically competitive. Can we afford this option? Can we afford not to consider this option? To have this energy avail-

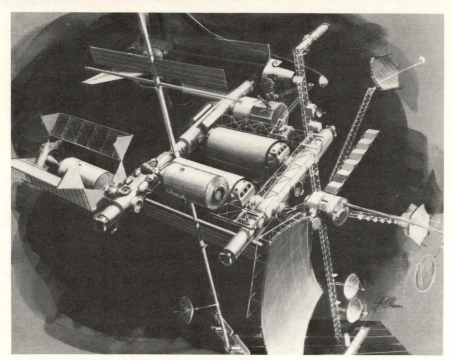

This designer's conception shows some of the applications of an advanced space operations center. This version of the spaceport shows the space shuttle unloading some of the modules which would comprise the system. These modules include living and command control quarters, warehouses for food, water, and hydrazine, and service areas containing batteries and other necessary supplies. Other areas of this advanced concept include hangars for spacecraft, solar panels to provide power for the station, and construction equipment to handle large structures. The large unit containing several antenna reflectors is a communications platform that is about to be attached to an orbital transfer vehicle for a flight to a higher orbit in space. (*NASA*)

able in 2030 requires a decision date of 1990. Much additional work must be done before a financial commitment of this magnitude can be reached.

Strategic Minerals

Figure 13-7 shows a list of strategic minerals upon which modern society depends. They are in short supply on earth and subject to the same constraints that OPEC has placed on petroleum. Are these minerals available in space? Today, no one knows but the moon's surface contains many useful minerals and large numbers of mineral-rich asteroids are in near-earth orbits. Our planet is, after all, an island in the sky, a spaceship on which we live with limited supplies of food, water, and minerals. As the limited

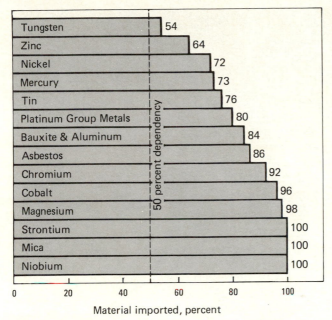

FIGURE 13.7
The share of strategic materials that the United States gets
from abroad [7].

amounts of strategic minerals on earth are used up, will modern society
decay or will wars be started to obtain these minerals? Or will ways be found
to mine minerals in space and return them to earth?

National Defense

One of the most important aspects of space is the role it plays in national
defense. About one-third of the early shuttle launches will be devoted to
Department of Defense payloads. The Russian space station *Salyut 7* has
been in orbit for over 4 years. The U.S.S.R. has been launching three times
as often as the United States for the past 5 years and most of these launches
have been militarily oriented. These include satellites for communication
and surveillance. (Satellite surveillance by both the Russians and Americans
are part of the open-sky policy established by President Eisenhower to
increase the chances for peaceful coexistence by preventing either side
from achieving a secret arms buildup. We need to know if it is possible for
one side to gain military superiority by deploying in space high-technology
weapons such as lasers. It is important to know what can be done in space
and also how to defend ourselves against an attack from space.

A small transportation vehicle waits for lift-off from a mining town on the moon. Lunar materials will be the prime resource for metals and oxygen for a proposed space colony. To the right of the transport are the mines and living quarters for the town. The smaller tube bins with glass windows are agriculture sheds where food for the town will be grown. To the left, lunar materials speed down a magnetic track. This system accelerates them to speeds that will allow them to escape the moon's gravity (one-sixth of the earth's gravity) and hurtle into space where they will be collected for construction of the proposed space colony. (*NASA*)

Commercial Uses

So far the biggest commercial success in space has been in the field of communications. Worldwide television and telephone transmissions are now routinely handled by satellites. This happened because communication satellites are a cheaper and more reliable way to handle worldwide information transfers. The door is just beginning to open on other possible commercial uses of space. Observation from space will increase in importance to monitor crops, to explore for minerals, and to locate ocean resources. The low-gravity environment in space permits large crystals to be grown, which can improve the efficiency of lasers. Chemicals can be mixed with much higher levels of purity, a practice which will advance medicine. As we become more familiar with space, engineers and scientists will think of many ways to advance the quality of life by using space and its resources.

Engineers in Space

Figure 13-8 suggests some possible ways engineers will be used in the space program. Some will be ground-based, while others will work and perhaps live in space.

What, then, is the challenge to engineers? Will this nation, the descendants of those who explored and settled the new world called America, the nation that placed the first man on the moon, sit by and watch others pioneer the next frontier? Will this country's leaders become so engrossed in domestic matters that no further space goals will be established? Time will tell. However, when the day arrives that this country again makes a commitment to space exploration, many of the space pioneers will be men and

FIGURE 13.8
The role of the engineer in space programs.

AERONAUTICAL AND ASTRONAUTICAL
ENGINEERS

- Analyze aerodynamics
- Wind-tunnel testing
- Transportation
- Propulsion systems
- Structures
- Control systems

AGRICULTURAL ENGINEERS

- Earth resources payloads

BIOMEDICAL ENGINEERS

- Environmental control systems
- Biological experiments

CHEMICAL ENGINEERS

- Propulsion systems
- Thermal protection systems
- Manufacturing payloads

CIVIL ENGINEERS

- Transportation
- Structures

ELECTRICAL ENGINEERS

- Electrical power systems
- Computer systems
- Communication systems
- Payloads

INDUSTRIAL ENGINEERS

- Ground facilities
- Manufacturing payloads

MATERIALS ENGINEERS

- Thermal protection systems
- Propulsion systems

MECHANICAL ENGINEERS

- Payloads
- Mechanical systems
- Control systems

NUCLEAR ENGINEERS

- Payload power supplies

women engineers like you. As some European nations—Spain, Portugal, and England—benefited greatly from the exploration of America, so too will some nations benefit from space exploration, a high-technology area of the future.

PRODUCTIVITY

"Productivity" has become a popular word that businessmen, economists, and politicians have been throwing around frequently in recent years. What does productivity mean? Why is it important? This section will define two aspects of productivity, explain why they are important, discuss some of the factors that affect productivity, and point out one factor that is the key to winning the productivity race. (Can you guess what that key factor is?)

Two Definitions

When speaking of productivity one must be careful to recognize the difference between productivity and rate of productivity growth. If you are in a

Productivity can be increased by turning repetitious, dangerous jobs over to computer-controlled robots. This robotic welding line is programmed to apply nearly 3000 welds to each body cycled through the system. (*Chrysler Corporation*)

race your *speed* (productivity) is important; equally important is how fast your competitors are *accelerating* (rate of productivity growth). These two parameters tend to get mixed up when productivity is discussed.

Productivity is generally defined as the ratio of goods and services produced to all the inputs needed to produce these goods and services. Inputs include factors such as people, machines, raw materials, research, and development. Hence,

$$\text{Productivity} = \frac{\text{output}}{\text{input}}$$

This ratio is used to measure the efficiency of a single business operation or the efficiency of the complete business economy of a nation, as shown in Figure 13-9. This ratio then measures the speed of a nation in the industrial race.

As can be seen in Figure 13-9, you are living in the most productive country in the world. The United States has automated and computerized manufacturing lines that operate without the manual aid of a single person and has the most productive farmlands in the world. The United States produces more goods and services per worker than any other nation in the world. However, two critical factors stand out when the United States posi-

FIGURE 13.9
Comparison of the total productivity of free-world nations in 1977 to that of the United States (100 percent). (*American Productivity Center*)

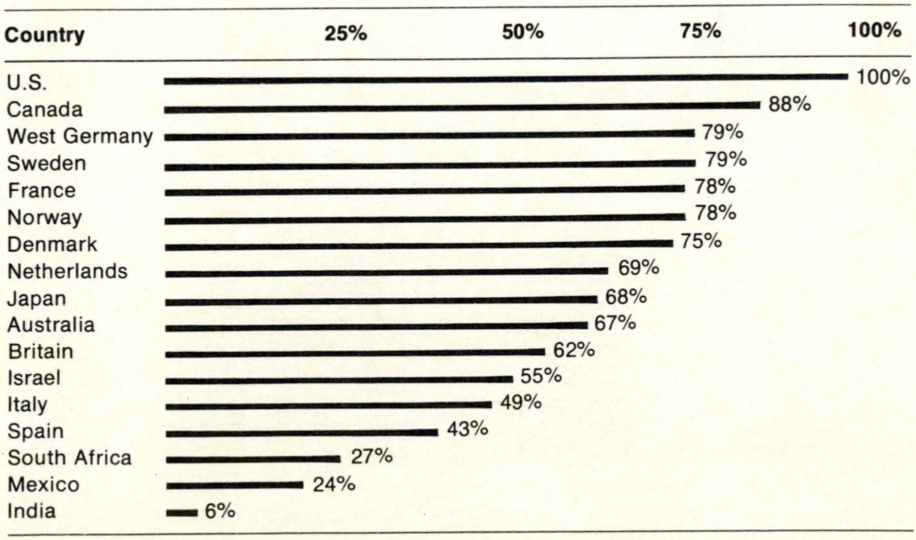

Country	25%	50%	75%	100%
U.S.				100%
Canada				88%
West Germany				79%
Sweden				79%
France				78%
Norway				78%
Denmark				75%
Netherlands			69%	
Japan			68%	
Australia			67%	
Britain			62%	
Israel		55%		
Italy		49%		
Spain		43%		
South Africa	27%			
Mexico	24%			
India	6%			

tion is examined from the productivity standpoint. As shown in Figure 13-10, although productivity is high, productivity growth has declined substantially in the past few years.

Rate of productivity growth (acceleration) is a measure of the factors that will determine future productivity (speed) of a business or a nation. Figure 13-10 compares the percent annual change in productivity of some of the leading industrial nations. One can easily see that while the United States presently is the most productive nation in the world, as shown in Figure 13-9, it also has had one of the lowest rates of growth of productivity for the past 30 years. This has very serious implications for the economic health and well-being of our nation.

For 25 years after World War II, the United States exhibited an average annual growth of 3 percent. Since 1973, the United States has shown an annual growth rate of only 0.2 percent. And even more frightening, in 1977 through 1979, our growth rate has been negative. While our productivity growth has been decreasing, other major industrial nations have shown rapid productivity increases. For instance, Japan's annual growth rate has averaged almost 6 percent per year for the past 10 years. Declining American productivity growth and a declining position among other industrial countries are cause for major concern.

Why Productivity Is Important

Figure 13-9 shows the relative percentage of seventeen free-world countries in relation to total productivity of the United States. There are a number of

FIGURE 13.10
Comparative data on productivity in terms of real domestic product per employed person, 1950–1978. (*U.S. Bureau of Labor Statistics*)

	Relative productivity[1] 1978	Average annual percent change in productivity[2]		
		1950–1965	1965–1973	1973–1978
Japan	63.0	7.2	9.1	3.1
West Germany	85.1	5.2	4.3	3.2
Italy	57.3	5.1	5.6	1.3
France	85.6	4.7	4.5	2.8
Canada	96.1	2.7	2.3	0.8
United Kingdom	58.4	2.2	3.3	0.9
United States	100.0	2.4	1.6	0.4

[1] Real gross domestic product per employed person using international price weights, relative to the United States.

[2] Growth in real domestic product per employed person using own country's price weights.

advantages to living in countries like the United States and Canada, with high productivity, compared with living in countries with low productivity. These advantages include:

1 Money available to educate our young people (your tuition covers less than half the cost of your college education)
2 Better health standards including lower infant mortality rates and longer life expectancy
3 Higher standards of living that permit us to live more comfortably, take vacations, and have money left over to help others

These advantages, and many others that we enjoy and others seek to achieve, are a result of living in a highly productive nation.

Higher productivity means increased ability to compete in the marketplace. This means larger profits, larger capabilities to invest (in research, new plants, and equipment), higher wages, higher employment, larger dividends for stockholders, and/or lower prices. However, nations and industries must maintain or improve their productivity relative to their competitors to stay in business. Therefore, the rate of change of productivity is equally important. As the productivity of a country increases, gross national product (GNP) increases since more output is produced per employee.

If a country experiences a decrease in productivity, the standard of living and economy suffer. The country is less able to compete in the ever-growing world market. Industry is less able to invest in new equipment and plants, less able to develop new products, and less able to control their product costs. Wages are lower and quality suffers. Natural resources are not utilized efficiently. A low-productivity country generally becomes more dependent on productive nations in the world for goods, services, and new technologies.

Factors That Affect Productivity

The relative importance of factors that affect productivity is sometimes debated, but there are several that are commonly agreed upon. These factors include the level of capital investment in business, utilization of plant capacity, levels of research and development, governmental regulations and taxation, economic factors such as energy costs, and the changing attitudes and lifestyle of the work force.

Factors that Affect Productivity Growth

Several of these factors have also contributed to the recent decline in American productivity growth.

Research and Development   One of the major causes is the slowdown in research and development spending. In a 10-year period following 1968, total R&D expenditures as a percent of GNP dropped from 3.0 to 2.2 percent. Business R&D expenditures and spending on basic research (that which results in future productivity improvement) have also declined significantly. Research and development is the basis for new product development.

Government Regulations As a result of governmental regulation and legislation, industry has spent billions of dollars in the areas of industrial health and safety, and in environmental protection. While these things are important, their costs do little to improve productivity or the competitiveness of goods in the marketplace. For instance, in 1976, at the peak of our government's drive to decrease pollution, about 8.6 percent of all manufacturing plant and equipment expenditures were for the purpose of pollution abatement.

Capital Investment Insufficient capital investment is another major contributing factor to the decline in productivity growth. Data in Figure 13-11 reveal that there is a strong positive correlation between capital investment and the growth rates of productivity and GNP of a nation. Since the 1960s, business investment in the United States as a percent of GNP has been well

FIGURE 13.11
Productivity improvement in relation to capital investment.
(*American Productivity Center*)

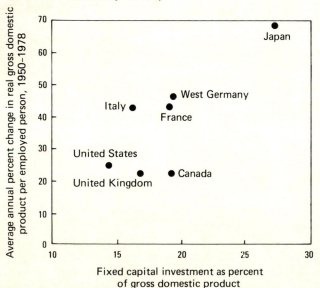

Computer-aided design and manufacturing (CAD/CAM) promises to make industry vastly more productive. One facet of this is computer graphics, a technology that frees the design engineer from the drawing board and the rigid constraints of making precise designs on paper. Instead, he can use the computer terminal to create much more complex designs, such as this three-dimensional teapot, in a fraction of the time and at a far lower cost than before. These designs can be instantly altered or rotated to expose a different view at the push of a button. A major effort is underway to extend the capabilities of this powerful new design technology. (*General Electric Company*)

below that of other industrialized countries. Many experts blame the United States tax system which encourages consumption and discourages business investments in new plants and equipment for this negative effect on productivity.

Automation and Robotics   Capital investments that improve productivity include automation of overall manufacturing operations and the application of robotics. The use of computers and robots can reduce labor costs and increase productivity. Robots have particular value for simple, routine, monotonous jobs such as assembly-line painting and welding. They are also useful in job environments where health and safety are of particular concern. One robot can often do the work of five workers at a lower cost. Robot sales in the United States have been growing rapidly, with the automobile industry leading in the number of purchases. Japan is rapidly becoming known for its growing integration of robots in complex manufacturing processes. Relative to the size of their economy, Japan has purchased twice as many robots as the United States. The application of robotics has considerable potential to boost the productivity in the United States.

Other Factors   In the past decade, economic instabilities such as rising inflation, recession, and an atmosphere of uncertainty have been detrimental to productivity. Other factors include the recent rise in energy costs, a changing American work ethic, an influx of inexperienced workers within the labor force, short-sighted management, and a dwindling supply of natural resources.

The Key Factor in Productivity and Productivity Growth

Probably the most important factor that will determine our success in improving productivity was discussed in Chapter 7. No matter how efficiently we produce gas guzzling cars or buggy whips it will not help our productivity. The keystone to a successful drive for improved productivity will be the creativeness of our people, particularly our engineers. The ability to develop new products such as computer chips, new technologies such as satellite communications, and new products like lasers may lead to the breakthrough needed to continue the success of our free enterprise system.

Future Outlook

Economic forecasts of worldwide productivity growths are not encouraging for the United States. As shown in Figure 13-12, if current trends persist, Germany, France, and Japan will surpass the U.S. productivity levels in 1984, 1985, and 1992, respectively.

Summary

The declining trend in United States productivity growth can be reversed with the cooperation of business, government, and private citizens. Engineers are faced with the challenge to improve American productivity. Whether employed as production, construction, or design engineers, researchers, or corporate managers, engineers will be involved with pro-

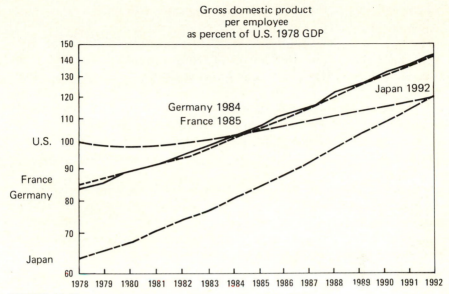

FIGURE 13.12
Projected international productivity trends. (*American Productivity Center*)

ductivity improvement in one way or another. Regaining leadership will require creativity, ingenuity, technical competence, and financial wisdom. This represents a great challenge—the challenge to improve the strength and well-being of this country and its citizens.

Conclusions

In this chapter we have briefly discussed energy, pollution, space, and productivity, four of the key challenges to the engineering profession. These subjects are very complex and there are no easy answers to the problems they present. Technical solutions are available to many of these problems but a social consensus is needed before action will be taken. For society to reach technically feasible and economically reasonable decisions, the technical community must participate in the debate and decision-making process. You and your classmates must be prepared to do this.

REFERENCES

Energy

1 Legassie, R. W. A., and F. I. Ordway, "A Quick Look at the National Energy Plan," *Astronautics and Aeronautics*, November 1977.

2 Bueche, A. M., "The Hard Truth About Our Energy Future," General Electric Company, PRD-459, June 1979.
3 "Lights Out by Year 2000 Without Increased Coal, Nuclear Energy Production," *Engineering Education*, vol. 7, no. 9, March 1981.
4 "The Need for Change, the Legacy of TMI," *Report of the President's Commission on the Accident at Three-Mile Island*, Pergamon, New York, 1979.

Pollution

5 Carey, Irene S, "The States and the Environment: Nine Governors Tell What They Think Needs To Be Done Next," *Context*, vol. 6 no. 2, E.I. duPont de Nemours & Co., 1977.
6 Jastrow, Robert, "Weather Forecast: The Lull Before the Storm," *Family Week*, March 15, 1981.

Space

7 Bluth, B. J., and S. R. McNeal (eds.), *Update on Space*, vol. 1, National Behavior Systems, Granada Hills, Calif., 1981.
8 Cook, James, "The Revolution in Space," *Forbes*, February 15, 1982, p. 90.

Productivity

9 American Productivity Center, "Productivity Perspectives, A Chartbook of Key Facts on U. S. Productivity in an Increasingly Competitive World," 1980.
10 Heer, Ewald, "Robots in Modern Industry," *Astronautics and Aeronautics*, September 1981.

ASSIGNMENTS

1 Write a four- to six-page report on one type of energy mentioned in this chapter, and its specific relation to the engineer. Use the *Reader's Guide to Periodical Literature* to obtain up-to-date references. Follow the rules of report writing outlined in Chapter 11.
2 Write a four- to six-page report on one type of pollution mentioned in this chapter, and its specific relation to the engineer. Use the *Reader's Guide to Periodical Literature* to obtain up-to-date references. Follow the rules of report writing outlined in Chapter 11.
3 Write a four- to six-page report on some engineering aspect of space. Use the *Reader's Guide to Periodical Literature* to obtain up-to-date references. Follow the rules of report writing outlined in Chapter 11.
4 Write a four- to six-page report on a productivity device of interest to engineers. Use the *Reader's Guide to Periodical Literature* to obtain up-to-date references. Follow the rules of report writing outlined in Chapter 11.
5 Are robots the answer to solving the U.S. productivity problem? Where can they be most effectively used by industry? Where are they not useful? Will the industrial

robot put the American worker and the engineer out of work? Discuss these questions by writing a four- to six-page report.

6 Some experts feel that water will be a major cause of concern in the United States before the end of this century. What water problems exist in your area, in the southwest, and in New England? Do you expect them to get better or worse in the next 10 years? Discuss in a four- to six-page report.

The characteristics of engineering
   Engineering as a demanding and rewarding career
Your qualifications for engineering

14

WHO SHOULD
STUDY ENGINEERING

The primary objective of this book has been to provide the information about engineering needed by someone contemplating a career in this field. If the authors have accomplished their purpose, you should know what engineering is, what engineers do, and how they go about doing it. You should also have learned something about your own interests and abilities from your reaction to the reading material and the assignments. The third step in career selection is a comparison of your desires with what engineering has to offer and a check of your qualifications with those required for success in engineering.

CONCLUSIONS ABOUT ENGINEERING AS A CAREER

To help you in your analysis, let us first list some broad conclusions that can be drawn from previous discussions and then raise some questions that you alone can answer.

Engineering Is Important Work in This Technological Age

Engineers provide a vital link between scientists and the general public. Their raw material consists of new discoveries, new ideas, new knowledge;

it is their job to convert scientific advances to practical forms, produce them economically, and operate them efficiently. As a result of the engineer's efforts, every industrial worker, business executive, and homemaker has command over many mechanical slaves which have, to a large extent, freed human beings from drudgery and toil.

The results of engineering affect everyone on earth. Engineers contribute to the health, standard of living, and general welfare of our nation by developing new sources of energy, increased supplies of food, and improved means for transportation and communication. In nations lacking engineering leadership, the technology gap is clearly evident.

Engineering Includes a Wide Spectrum of Functions

Of particular interest to prospective students is the fact that engineering is not a single activity; instead it embraces a wide variety of functions with rather distinctive features. In order of decreasing emphasis on scientific technology, the major functions are research, development, design, construction, production, operation and maintenance, application and sales, and management.

For work in research, development, and design, engineers must have intensive training in science and mathematics. Graduate work is highly desirable. Imagination and creativity in scientific aspects and skill in experimentation are important qualifications. On the other hand, knowledge of engineering practice and skill in handling economic factors are not so important.

In the functions of construction, production, and operation, the major emphasis is on the art of engineering—solving technological problems economically, efficiently, and on schedule. For work in these functions a feeling for what machines and people can do is important. There is need for skill in human relations and in solving personal problems of a type that may not enter into a research engineer's work.

Management and sales functions are similar in that human and economic factors are frequently more important than technological factors. Personal characteristics may be more important than scientific training. Problems tend to be more qualitative in nature than in other functions, and precise mathematical methods are less applicable. Sensitivity in human relations and understanding of basic human drives are necessary for success.

Engineering Is a Broad Field with Branches That Challenge Any Technical Interests

Engineering is an attitude or an approach rather than a specific subject. At one extreme engineers are concerned with colossal structures and tremendous energies; at the other end of the size scale they are solving the mysteries of molecular structure and the behavior of electrons. Some of their activities are in the realm of theoretical science; others depend primarily on practical experience. Some of the engineer's problems are purely technical and quantitative; others require a fine sense of human relations for their solution.

The engineer works with foods and fertilizer; penicillin and perfume; jet planes and baby carriages; ore crushers and dishwashers; radio waves, sound waves, and tidal waves; steam turbines and steam irons; solar furnaces and home refrigerators; steel cable and nylon thread. Within the broad branches of civil, mechanical, electrical, aerospace, industrial, materials, or chemical engineering or within the more specialized subfields, any technically minded person can find an area of interest.

Engineering Is Difficult Work Requiring Special Abilities

The distinguishing characteristic of engineering work is the emphasis on quantitative thinking—reasoning in mathematical terms and performing calculations with numbers and symbols. Engineers make use of advanced mathematical methods and abstract concepts that, for many people, are extremely difficult to comprehend, let alone apply. In addition, they must be able to express their ideas verbally and graphically.

The engineering approach to problem solution is based on rational thinking rather than emotional reaction. Logical, step-by-step analysis accompanied by careful checking and leading to well-founded conclusions must become habitual, but in the quest for optimum solutions, imagination and creativity are essential.

Engineers are doers as well as planners, and frequently they are responsible for the completion of an overall project involving many other people. They must be able to work long hours and under pressure in situations where every hour of delay means financial loss. Because engineering activities always involve the future, they must be willing to make decisions under conditions of uncertainty. Because their work may influence the lives of many people, they bear a special responsibility to anticipate any adverse effects.

The rapid rate of technological progress makes engineering education a lifetime project. After four or five years of intensive study, new graduates find that they have a knowledge of fundamentals but know little of the art of engineering. They must study at work and in their spare time to learn their particular jobs. To keep up with new developments they must study the technical journals and attend meetings and conferences. If they ever stop learning, they are quickly left behind and are of little use to their employers or their clients.

**Engineers Are Members of a Technical Team That Includes
Technologists and Scientists**

Technologists usually specialize in one aspect of engineering, and therefore, their training can be less extensive and more intensive. Working with engineers, they use mathematics and science in solving problems; they use precision instruments for making measurements, and they may use tools for construction. Technologists usually have a strong interest in the practical aspects of technology and less interest in scientific theory. They should have skill in manipulating apparatus and equipment along with the ability to handle elementary mathematics and science.

Scientists are also specialists—at an advanced level usually requiring graduate work. They are engaged in discovering new principles and learning new facts without being concerned with practical applications. They sometimes work with engineers, contributing their specialized knowledge to the solution of the engineering problems. Scientists need a high level of intelligence, ability in creative thinking, and skill in experimentation.

**Engineering Work Is Demanding, Satisfying,
and Rewarding**

The master's degree, increasingly expected for professional work, represents 5 years of intensive study. Although engineering study is difficult and requires long hours of concentrated application, achieving mastery of sophisticated concepts is highly satisfying.

Engineers are engaged in creative, productive, and constructive work. Whereas the doctor's patients are usually ill and the lawyer's clients are in trouble, the engineer is primarily concerned with healthy, growing, and productive activities. They have their problems with sick processes and production bottlenecks, of course, but for the most part their work is positive—involving advances, improvements, and benefits. Successful engineers know the joy of creation—creation of a new concept, device, process, or procedure out of their own knowledge, thought, and experience.

There is an expanding demand for creative engineers. Every foreseeable technological development will increase the need; automation, new energy sources, pollution control, space travel, weather control, higher productivity, city planning, advances in transportation and communication—all will require more technically trained persons and fewer laborers. Starting salaries for engineering graduates are relatively high, and for qualified engineers there are many opportunities for increased responsibility and compensation. While public acclaim comes to few members of the engineering profession, one's colleagues are quick to recognize professional competence and to honor outstanding achievement.

YOUR QUALIFICATIONS FOR ENGINEERING

Success in engineering requires a combination of interest, aptitude, and drive.

Are You Interested in Engineering?

Over a period of 35 years the authors have been questioning groups of graduating seniors regarding their employment goals; almost invariably, at the top of the list they put "interesting work." Interesting work arouses your curiosity and commands your attention. Where do your interests lie?

Look back over your life and see what ideas, things, and activities commanded your attention or aroused your curiosity. Have you always been interested in mechanical or electrical devices? What do you do in your spare time when you can do just what you like? Do your hobbies indicate an interest in technical things or abstract ideas? Are you fascinated by descriptions of new technical or scientific developments? Do you like mathematical puzzles? Would you like to work in a field where a considerable portion of your time would be spent in solving mathematical problems?

Which parts of the text did you find most interesting? Have you ever on your own initiative visited a plant or refinery or television station or construction project? Have you ever sought out engineers and questioned them about their work? Do you have sufficient interest in engineering to select it as a lifetime profession?

Are Your Aptitudes in line with the Requirements for Engineering?

On the basis of many observations as well as carefully devised tests, engineering educators agree that ability in mathematics is most closely associated with success in engineering study. How good are you at reasoning in quantitative terms? Can you manipulate abstract symbols without confusion? Would you rather read about automobile engines or make calculations on their efficiency? Do you grasp mathematical relations quickly, or do you need lots of time for mastery? How did your high-school teachers rate your ability in mathematics?

Next to mathematics, physics is perhaps the best indicator of your probable success in engineering study. Are you able to read the statement of a physical principle such as Newton's law of gravitation, understand the relationship among the variables, and visualize how the principle is applied in practice? Do abstract concepts such as acceleration, absolute zero temperature, and magnetic flux seem clear and easy to understand? Were you able to solve the problems involving applications of physical principles?

There are other aptitudes that will contribute to success in engineering, but they are probably secondary in importance; they are modifying factors rather than primary determining factors. Are you able to translate ideas expressed in words into clear sketches? Are you able to describe in words a complex motion such as the action of the valves in an automobile engine? How well do you get along with people? Can you persuade others to your way of thinking? Can you get a group of people to work together efficiently? Can you provide leadership that others are willing to follow?

Are you always considering how you might change things for the better? Are you unhappy when something doesn't work well or when a slight modification would greatly improve its performance? Can you discover what is impairing its operation and can you visualize a practical improvement?

Note that none of these qualifications is sex-related. Women are just as capable as men in the attributes that count in engineering. The clearly visible opportunities for challenging responsibilities and economic rewards make engineering one of the most attractive professions for women today.

Do You Have the Drive Necessary for Engineering Study and Practice?

Engineering study is not only difficult, but it is also exacting. There is so much to learn and so little time that many hours of concentrated and efficient study are required. Because so much of engineering is quantitative, a high degree of accuracy is necessary in measurements, observations, and calculations. "Almost right" or "correct except for the decimal point" is not acceptable.

How are you at getting things done? Are you in the habit of carrying assigned projects all the way through to completion? On time? Are you a self-starter, setting your own tasks and completing them? On difficult assignments in the technical portion of this book, were you willing to work a little bit harder? When you have made up your mind to do something, do obstacles serve as a challenge in bringing out your best efforts? Do you have the drive required to complete an engineering curriculum and gain the necessary preprofessional experience?

Success in engineering usually requires interest, aptitude, and drive; however, a deficiency in one characteristic can be compensated for by extra strength in the other two. There are numerous examples of competent engineers who were low on aptitude but made up for it in interest and drive. A student with high aptitude but low interest may find that participation in advanced courses involving practical applications develops the necessary interest.

For the qualified person with technical interests, engineering is a wonderful field full of a host of opportunities for exciting, satisfying work and a lifetime career. Good luck!

APPENDIX A

OTHER ENGINEERING AND RELATED SOCIETIES

OTHER ENGINEERING, SCIENCE, AND RELATED SOCIETIES AND ORGANIZATIONS (SEE ALSO TABLE 4-1)

Abbre-viation	Name of society	Date founded	No. of members
AAAS	American Association for the Advancement of Science	1848	113,800
AACE	American Association of Cost Engineers	1956	3,700
AAEE	American Academy of Environmental Engineers	1955	1,700
AAPG	American Association of Petroleum Geologists	1917	14,100
ACM	Association for Computing Machinery	1947	30,000
ACS	American Chemical Society	1876	115,000
ACS	The American Ceramic Society	1898	6,100
ACSM	American Congress on Surveying and Mapping	1941	10,200
AEA	American Engineering Association	1979	1,000
AEE	Association of Energy Engineers	1977	3,000
AEMS	American Engineering Model Society	1970	502
AES	Audio Engineering Society	1948	NIM
AFIPS	American Federation of Information Processing Society	1961	NIM

Abbre-viation	Name of society	Date founded	No. of members
AFS	American Foundrymen's Society	1896	NIM
AIA	American Institute of Architects	1857	29,000
AICAE	American Indian Council of Architects and Engineers	1974	200
AICP	American Institute of Certified Planners	1917	10,000
AIP	American Institute of Physics	1931	55,000
AIPE	American Institute of Plant Engineers	1954	8,100
AISE	Association of Iron and Steel Engineers	1907	2,500
AISES	American Indian Science and Engineering Society	1977	80
AMS	American Mathematical Society	1888	16,000
APCA	Air Pollution Control Association	1907	5,600
API	American Petroleum Institute	1919	7,000
APS	American Physical Society	1899	26,600
APWA	American Public Works Association	1894	¡6,700
AREA	American Railway Engineering Association	1899	3,500
ASA	Acoustical Society of America	1929	4,800
ASHRAE	American Society of Heating, Refrigerating and Air-Conditioning Engineers	1894	30,000
ASM	American Society for Metals	1913	9,000
ASNE	American Society of Naval Engineers	1888	4,100
ASP	American Society of Photogrammetry	1934	6,400
ASPEP	Association of Scientists and Professional Engineering Personnel	1945	1,600
ASQC	American Society for Quality Control	1946	29,000
ASSE	American Society of Sanitary Engineers	1906	2,800
ASTM	American Society for Testing Materials	1898	24,500
ATEA	American Technical Education Association	1928	1,500
DPMA	Data Processing Management Association	1951	22,500
ESL	Engineering Societies Library	1913	420,000
ICET	Institute for the Certification of Engineering Technicians	1961	60,000
IES	Illuminating Engineering Society of North America	1906	9,000
ISA	Instrument Society of America	1945	20,600
JETS	Junior Engineering Technical Society	1950	5,000
MAA	Mathematical Association of America	1915	18,700
MAES	Mexican American Engineering Society	1974	200
NABCE	National Association of Black Consulting Engineers	1976	NIM
NACE	National Association of Corrosion Engineers	1945	10,300
NAPE	National Association of Power Engineers	1882	11,000
NAS	National Academy of Science	1863	1,300
NSBE	National Society of Black Engineers	1975	1,500
NICE	National Institute of Ceramic Engineers	1938	1,600
NSBE	National Society of Black Engineers	1975	1,500
NTA	National Technical Association	1929	600
ORSA	Operations Research Society of America	1952	5,500

Abbre-viation	Name of society	Date founded	No. of members
SAE	Society of Automotive Engineers	1905	30,100
SAM	Society for Advancement of Management	1912	3,000
SAME	Society of American Military Engineers	1920	28,800
SAWE	Society of Allied Weight Engineers	1939	1,300
SBE	Society of Broadcast Engineers	1963	3,500
SES	Society of Engineering Science	1963	400
SFPE	Society of Fire Protection Engineers	1950	2,900
SHPE	Society of Hispanic Professional Engineers	1974	250
SMC	Science Manpower Commission	1953	20
SME	Society of Manufacturing Enineers	1932	43,800
SME	Society of Mining Engineers of AIME	1871	20,700
SNAME	Society of Naval Architects and Marine Engineers	1893	10,700
SPE	Society of Petroleum Engineers	1913	27,300
SPE	Society of Plastics Engineers	1941	20,200
SPHE	Society of Packaging and Handling Engineers	1945	2,000
SX	Sigma XI, The Science Research Society	1893	120,000
TMS	The Metallurgical Society of AIME	1957	8,000
UATI	Union of International Engineering Organizations	1951	NIM
UET	United Engineering Trustees	1904	NIM
UPADI	Pan American Federation of Engineering Societies	1949	NIM

Source: *Directory of Engineering Societies and Related Organizations,* American Association of Engineering Societies, 1979, New York, N.Y.
NIM = No individual membership.

APPENDIX B

ENGINEERING AND RESEARCH HONOR SOCIETIES

COLLEGE ENGINEERING AND RESEARCH HONORARY SOCIETIES

Name/Address	Branch of engineering	Year begun	Number of members	Chapters
Alpha Epsilon University of Missouri Columbia, MO 65211	Agricultural	1959	2,432	24
Alpha Pi Mu Industrial Engineering Wichita State University Wichita, KS 67208	Industrial	1949	4,421	57
Alpha Sigma Mu Materials Engineering Virginia Polytechnic Blacksburg, VA	Metallurgical and materials	1932	5,000	25
Chi Epsilon Civil Engineering University of Tennessee Knoxville, TN 37916	Civil	1922	44,000	91

Name/Address	Branch of engineering	Year begun	Number of members	Chapters
Eta Kappa Nu University of Illinois Urbana, IL 61081	Electrical	1904	110,000	145
Omega Chi Epsilon McNeese State University Lake Charles, LA 70609	Chemical	1931	7,000	35
Pi Tau Sigma Tennessee Technological University Cookeville, TN 38501	Mechanical	1915	59,779	115
Sigma Gamma Tau Aeronautical Engineering Purdue University W. Lafayette, IN 47907	Aerospace	1953	9,200	30
Tau Beta Pi Box 8840 University Station Knoxville, TN 37916	All branches of engineering	1885	230,000	51

Source: Association of College Honor Societies, "Booklet of Information," Library of Congress 76-8325, 1980–81, Williamsport, Pa.

APPENDIX C

MODEL RULES OF PROFESSIONAL CONDUCT

MODEL RULES OF PROFESSIONAL CONDUCT, A GUIDE FOR USE BY REGISTRATION BOARDS, NATIONAL COUNCIL OF ENGINEERING EXAMINERS, AUGUST 1979

Preamble.   In order to safeguard life, health and property, to promote the public welfare, and to establish and maintain a high standard of integrity and practice, the following Rules of Professional Conduct shall be binding on every person holding a certificate of registration and on all partnerships or corporations or other legal entities authorized to offer or perform engineering or land surveying services in this state.

The Rules of Professional Conduct as promulgated herein are an exercise of the police power vested in the Board by virtue of the acts of the legislature.

All persons registered under State Registration Laws are charged with having knowledge of the existence of these Rules of Professional Conduct, and shall be deemed to be familiar with their provisions and to understand them. Such knowledge shall encompass the understanding that the practice of engineering and land surveying is a privilege, as opposed to a right.

In these Rules of Professional Conduct, the word "registrant" shall mean any person holding a license or certificate issued by this board.

Fundamental Canons.   In the fulfillment of their professional duties

I.   Registrants shall hold paramount the safety, health and welfare of the public in the performance of their professional duties.

   a.   Registrants shall at all times recognize that their primary obligation is to protect the safety, health, property and welfare of the public. If their professional judgment is overruled under circumstances where the safety, health, property or welfare of the public are endangered, they shall notify their employer or client and such other authority as may be appropriate.

   b.   Registrants shall approve and seal only those design documents and surveys which are safe for public health, property and welfare in conformity with accepted engineering and land surveying standards.

   c.   Registrants shall not reveal facts, data or information obtained in a professional capacity without the prior consent of the client, or employer except as authorized or required by law.

   d.   Registrants shall not permit the use of their name or firm name nor associate in business ventures with any person or firm which they have reason to believe is engaging in fraudulent or dishonest business or professional practices.

   e.   Registrants having knowledge of any alleged violation of any of these rules of professional conduct, shall cooperate with the Board in furnishing such information or assistance as may be required.

II.   Registrants shall perform services only in the areas of their competence.

   a.   Registrants shall undertake assignments only when qualified by education or experience in the specified technical fields of engineering or land surveying involved.

   b.   Registrants shall not affix their signatures or seals to any plans or documents dealing with subject matter in which they lack competence, nor to any such plan or document not prepared under their direction and control.

   c.   Registrants may accept an assignment outside of their fields of competence to the extent that their services are restricted to those phases of the project in which they are qualified, and to the extent that they are satisfied that all other phases of such project will be performed by registered or otherwise qualified associates, consultants, or employees, in which case they may then sign and seal the documents for the total project.

   d.   In the event a question arises as to the competence of a registrant in a specific technical field which cannot be otherwise resolved to the State Board's satisfaction; the State Board, either upon request of the registrant or on its own volition, shall admit the registrant to an appropriate examination.

III.   Registrants shall issue public statements only in an objective and truthful manner.

   a.   Registrants shall be objective and truthful in professional reports, statements or testimony. They shall include all relevant and pertinent information in such reports, statements or testimony.

   b.   Registrants may express publicly a professional opinion on technical subjects only when that opinion is founded upon adequate knowledge of the facts and competence in the subject matter.

     c. Registrants shall issue no statements, criticisms or arguments on technical matters which are inspired or paid for by interested parties, unless the registrants have prefaced their comments by explicitly identifying the interested parties on whose behalf they are speaking, and revealing the existence of any interest the registrants may have in the matter.

IV. Registrants shall act in professional matters for each employer or client as faithful agents or trustees, and shall avoid conflicts of interest.

     a. Registrants shall disclose all known or potential conflicts of interest to their employers or clients by promptly informing them of any business association, interest, or other circumstances which could influence their judgment or the quality of their services.

     b. Registrants shall not accept compensation, financial or otherwise, from more than one party for services on the same project, or for services pertaining to the same project, unless the circumstances are fully disclosed to, and agreed to, by all interested parties.

     c. Registrants shall not solicit or accept financial or other valuable consideration, directly or indirectly, from contractors, their agents, or other parties in connection with work for employers or clients for which the registrant is responsible.

     d. Registrants in public service as members, advisors or employees of a governmental body or department shall not participate in decisions with respect to professional services solicited or provided by them or their organizations.

     e. Registrants shall not solicit or accept a professional contract from a governmental body on which a principal or officer of their organization serves as a member, except upon public disclosure of all pertinent facts and circumstances and consent of appropriate public authority.

V. Registrants shall avoid improper solicitation of professional employment.

     a. Registrants shall not falsify or permit misrepresentation of their, or their associates' academic or professional qualifications. They shall not misrepresent or exaggerate their degree of responsibility in or for the subject matter of prior assignments. Brochures or other presentations incident to the solicitation of employment shall not misrepresent pertinent facts concerning employers, employees, associates, joint ventures or past accomplishments with the intent and purpose of enhancing their qualifications and their work.

     b. Registrants shall not offer, give, solicit or receive, either directly or indirectly, any commission, or gift, or other valuable consideration in order to secure work, and shall not make any political contribution in an amount intended to influence the award of a contract by public authority, but which may be reasonably construed by the public of having the effect or intent to influence the award of a contract.

INDEX

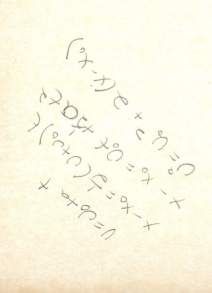